THEN and NOW

How the first 100 surgical patients admitted to the Massachusetts General Hospital were treated THEN (1821-1823) and how they would be treated NOW (2011)

Stephen P. Dretler, MD
Editor

THEN and NOW

The Massachusetts General Hospital (MGH) History Committee met in 2011 at the occasion of the bicentennial of the founding of the hospital in 1811. The committee knew that the records of the first one hundred patients admitted to the surgical service, beginning in 1821, were available.

The records of the first one hundred patients were found in the archives of the hospital and were cared for by MGH archivist, Jeff Mifflin. The records were pen and fading ink on brown writing paper with frayed and crumbling edges. Reading the nineteenth-century script on fragile paper was a significant challenge, not only extracting the information but also preventing damage to the records. Although there were many indecipherable words, there was only one record that was totally unintelligible.

Translating the flourished and flowery script of that time period (1821-1823) was tedious. For fear of damaging the records, all translation and case summaries were done in the archives of the hospital with wristwatches off and sleeves rolled up to prevent damage to the fragile paper. Once a week, two cases were reviewed and summarized.

After the case records were read, and pathological conditions and treatments were reviewed and written up, the cases were labeled THEN, referring to the understanding of the patient's disease and treatment at the time.

Department chairs at the MGH were then asked to have their associates write up summaries of how we now understand the disease process and what treatments we would give if the patients were treated today, in 2011. Thus the title of this work is "THEN and NOW."

My deepest gratitude goes to those who took time from their busy schedules to read cases and explain to us how these conditions are now treated, two hundred years later.

List of Cases ("THEN") and Commentaries ("NOW")

THEN - Patient Number	NOW - Commentary
1. Hemorrhoids	Dr. Paul C. Shellito
2. Urethral foreign body	Dr. Stephen Dretler
3. Popliteal aneurysm	Dr. Ashby Moncure
4. Urethral stricture	Dr. Stephen Dretler
5. Anal fistula	Dr. Paul C. Shellito
6. Gonorrhea	Dr. Dianne Sacco
7. Closed long-bone fracture	Dr. Stephen T. Gardner Dr. Harry E. Rubash
8. Hip dislocation	Dr. James Herndon Dr. R. Malcolm Smith Dr. Mark Vrahas
9. Chronic blepharitis	Dr. James Chodosh Dr. Matthew F. Gardiner Dr. Joan W. Miller Dr. Sotiria Palioura Dr. Athanasios Papakostas
10. Inflamed testicle	Dr. Adam Feldman
11. Vertebral trauma	Dr. Joseph Schwab
12. Compound femoral fracture	Dr. Jeffrey R. Jaglowski Dr. Harry E. Rubash
13. Urethral stricture	Dr. Stephen Dretler
14. Compound tibial fracture	Dr. Jeffrey R. Jaglowski Dr. Harry E. Rubash

15. Venereal disease — Dr. Esther Freeman

16. Frostbite of toes — Dr. Michael Watkins

17. Venereal disease — Dr. Esther Freeman

18. Cellulitis of toe — Dr. George Velmahos

19. Open oblique leg fracture — Dr. Jeffrey R. Jaglowski
    Dr. Harry E. Rubash

20. Stab wound of leg — Surgical Service

21. Penetrating injury of skull — Dr. Ann-Christine Duhalme

22. Urethral stricture — Dr. Stephen Dretler

23. Corneal abrasion — Dr. James Chodosh
    Dr. Matthew F. Gardiner
    Dr. Joan W. Miller
    Dr. Sotiria Palioura
    Dr. Athanasios Papakostas

24. Cancer of the breast — Dr. Barbara Lynn Smith

25. Compound leg fracture — Dr. Michael D. Baratz
    Dr. Harry E. Rubash

26. Death from head injury — Patient dead on arrival

27. Venereal ulcers — Dr. Esther Freeman

28. Elbow swelling — Minor orthopedic condition

29. Case missing — No commentary

30. Leg swelling — Dr. George Velmahos

31. Genital ulcers	Dr. Esther Freeman
32. Comminuted fracture of humerus	Dr. Stephen T. Gardner Dr. Harry E. Rubash
33. Ankle sprain	Minor orthopedic condition
34. Injury to skin of foot	Dr. George Velmahos
35. Musculoskeletal infection	Dr. Michael D. Baratz Dr. Harry E. Rubash
36. Urethral stricture	Dr. Stephen Dretler
37. Cancer of the breast	Dr. Barbara Lynn Smith
38. Case missing from archives	No commentary
39. Bilateral patella fracture	Dr. Jeffrey R. Jaglowski Dr. Harry E. Rubash
40. Muscular skeletal infection (psoas abscess)	Dr. Michael D. Baratz Dr. Harry E. Rubash
41. Squamous cell carcinoma of vagina	Dr. John O. Schorge
42. Lumbar herniated disc	Dr. Stephen T. Gardner Dr. Harry E. Rubash
43. Musculoskeletal inflammation	Dr. Michael D. Baratz Dr. Harry E. Rubash
44. Foreign body in eye	Dr. James Chodosh Dr. Matthew F. Gardiner Dr. Joan W. Miller Dr. Sotiria Palioura Dr. Athanasios Papakostas
45. Breast tumor	Dr. Barbara Lynn Smith

46. Sprained knee   Orthopedic Service

47. Parotid tumor   Dr. William G. Austen, Jr.

48. Hemorrhoids   Dr. Paul C. Shellito

49. Anal fistula   Dr. Paul C. Shellito

50. Migrating joint pain   Minor orthopedic condition

51. Cataracts   Dr. James Chodosh
                                           Dr. Matthew F. Gardiner
                                           Dr. Joan W. Miller
                                           Dr. Sotiria Palioura
                                           Dr. Athanasios Papakostas

52. Infected puncture wound of foot   Dr. George Velmahos

53. Open fracture of humerus   Dr. Michael D. Baratz
                                           Dr. Harry E. Rubash

54. Blunt trauma to finger   Minor orthopedic condition

55. Tumor of foot   No commentary

56. Rib abscess   Surgical Service

57. Uveitis   Dr. James Chodosh
                                           Dr. Matthew F. Gardiner
                                           Dr. Joan W. Miller
                                           Dr. Sotiria Palioura
                                           Dr. Athanasios Papakostas

58. Soft tissue injury   Dr. Eric M. Black
                                           Dr. Harry E. Rubash

59. Soft tissue mass (lipoma)   Dr. Keith Lillemoe

60. Soft tissue infection   Surgical Service

61. Contusion of knee — Minor condition

62. Venereal disease — Dr. Esther Freeman

63. Scapula bruise — Minor injury

64. Childhood cataracts — Dr. James Chodosh
Dr. Matthew F. Gardiner
Dr. Joan W. Miller
Dr. Sotiria Palioura
Dr. Athanasios Papakostas

65. Iliac aneurysm — Dr. R. Todd Lancaster
Dr. Virendra I. Patel

66. Trigeminal neuralgia — Dr. Leonard B. Kaban
Dr. David A. Keith
Dr. Thomas B. Dodson

67. Frostbite of toes — Dr. Michael Watkins

68. Tibial fracture — Dr. Stephen T. Gardner
Dr. Harry E. Rubash

69. Cataract — Dr. James Chodosh
Dr. Matthew F. Gardiner
Dr. Joan W. Miller
Dr. Sotiria Palioura
Dr. Athanasios Papakostas

70. Knee bruise — Minor orthopedic condition

71. Chronic loss of sight — Dr. James Chodosh
Dr. Matthew F. Gardiner
Dr. Joan W. Miller
Dr. Sotiria Palioura
Dr. Athanasios Papakostas

72. Syphilis — Dr. Esther Freeman

73. Migratory arthritis					Non-surgical condition

74. Musculoskeletal infection				Dr. Michael D. Baratz
							Dr. Harry E. Rubash

75. Osteomyelitis					Dr. Michael D. Baratz
							Dr. Harry E. Rubash

76. Child with residual cataract			Dr. James Chodosh
							Dr. Matthew F. Gardiner
							Dr. Joan W. Miller
							Dr. Sotiria Palioura
							Dr. Athanasios Papakostas

77. Hemorrhoids						Dr. Paul C. Shellito

78. Syphilis						Dr. Esther Freeman

79. Clinical case - unreadable				No commentary

80. Femoral fracture					Dr. James Herndon
							Dr. R. Malcolm Smith
							Dr. Mark Vrahas

81. Minor foot trauma					No commentary

82. Severe conjunctivitis				Dr. James Chodosh
							Dr. Matthew F. Gardiner
							Dr. Joan W. Miller
							Dr. Sotiria Palioura
							Dr. Athanasios Papakostas

83. Partially dissolved cataract			Dr. James Chodosh
							Dr. Matthew F. Gardiner
							Dr. Joan W. Miller
							Dr. Sotiria Palioura
							Dr. Athanasios Papakostas

84. Hand trauma						Minor injury

85. Breast mass — Dr. Barbara Lynn Smith

86. Pelvic mass — Gynecology Service

87. Leg ulcer/foot pain — Dr. Eric M. Black
Dr. Harry E. Rubash

88. Eye infection — Dr. James Chodosh
Dr. Matthew F. Gardiner
Dr. Joan W. Miller
Dr. Sotiria Palioura
Dr. Athanasios Papakostas

89. Syphilis — Dr. Esther Freeman

90. Eyelid scar — Dr. Joan W. Miller

91. Venereal disease — Dr. Esther Freeman

92. Cataracts — Dr. James Chodosh
Dr. Matthew F. Gardiner
Dr. Joan W. Miller
Dr. Sotiria Palioura
Dr. Athanasios Papakostas

93. Cataracts — Dr. James Chodosh
Dr. Matthew F. Gardiner
Dr. Joan W. Miller
Dr. Sotiria Palioura
Dr. Athanasios Papakostas

94. Carcinoma of the rectum — Dr. Paul C. Shellito

95. Soft tissue injury — Dr. Eric M. Black
Dr. Harry E. Rubash

96. Soft tissue inflammation — Minor orthopedic condition

97. Varicocele — Dr. Adam Feldman

98. Varicocele — Dr. Cigdem Tanrikut

99. Syphilis — Dr. Esther Freeman

100. Hemorrhoids — Dr. Paul C. Shellito

Patient #1: Hemorrhoids

THEN

The first surgical patient was admitted on September 20, 1821. He was an Irish-born seaman who had had hemorrhoids for eleven years. They had recently increased to the size of walnuts, causing pain with each step and blood with each bowel movement. He also had a rectal stricture three inches from the anus. Treatment consisted of ligation and excision of the hemorrhoids along with progressive, twice-daily anal dilation with bougies to the size of an index finger. The treatment was carried out for one month and the patient was discharged on October 20, 1821.

NOW

The first line of treatment for hemorrhoids in 2011 is conservative and stresses the use of stool softeners to reduce the strain necessary to produce a bowel movement and the use of ointments to reduce anal swelling. Sclerotherapy is used for smaller symptomatic hemorrhoids but is less effective than rubber band ligation, a treatment which cuts off circulation to the hemorrhoid and allows the lesion to wither and fall off. If bleeding is persistent from these thrombosed and dilated veins, surgery is performed. The surgical technique varies little from what was done "Then," but "Now" we have the benefit of local, regional, or general anesthetic. The proctoscope was introduced into clinical practice by The Welch Allyn Company in 1924. Without this instrument treatment was limited to external hemorrhoids

Dr. Paul C. Shellito

Patient #2: Urethral foreign body

THEN

A 20-year-old male was admitted September 28, 1821 with difficulty voiding. He had fallen the previous May and injured his urethra. Feeling an obstruction to his urine flow, he tried to dilate the narrowing with a "bean" of some sort. It appears that the foreign body became lodged in the urethra. Initially it was palpated by rectal exam. Subsequently the foreign body couldn't be felt and it was presumed to have fallen back into the bladder. A urethral sound was placed on October 16 and the stone was not felt. The bladder was filled with warm water, a string was tied around the urethra to prevent premature bladder emptying, and the patient was given tincture of opium and restrained in a "lithotomy" position. The sedative, laudanum, was given in place of the yet-undiscovered anesthesia. An operation was necessary to remove the foreign body, prevent obstruction, and relieve the pain. With the patient sedated and restrained, a knife was used to enter the perineum, trying to avoid both the prostate and the rectum. The introduction of the knife was followed by the operator's index finger to dilate the tract and locate the stone. A forceps was then used to extract the foreign body, which was indeed a bean. The patient experienced severe pain and the urine that filled the bladder came gushing out.

He was able to stand erect by November 30 and voided three times per urethra on December 1. His surgeon had successfully avoided the rectum and the urinary sphincter, and despite the lack of a sterile field, the patient did not become septic nor have significant bleeding (less than a pint in the first five hours). There is no note about the remaining urethral stricture, though the passing of the sounds probably dilated the narrowing.

NOW

If treated today, this patient would have general anesthesia and an endoscopic transurethral procedure, first using a guide wire and high pressure balloon to dilate the narrow urethra followed by a rigid cystoscope (19-21F) or a flexible cystoscope (14F) and a Holmium-Yag laser passed through the working channel of the endoscope. The stone or foreign body would be fragmented to particles which would be extracted, evacuated or, if small enough, left to pass spontaneously. Unless the patient were septic, it is likely that he would be discharged from the hospital the same day. If the foreign body was composed of material that did not absorb light the wavelength of the Holmium laser, grasping forceps would be passed per urethra, the foreign body crushed or morcellated, and the fragments extracted. If the object was too hard to break, the bladder would be filled, and under ultrasound view, a suprapubic puncture would be made and the tract dilated and pieces extracted. If a puncture fails, then a suprapubic incision would be made, the bladder entered, and the foreign body extracted. The suprapubic cystotomy would be followed by placement of a urethral catheter for drainage for a period of seven days, the puncture checked for leakage by a cystogram, and if no extravasation, the catheter would be removed.

Dr. Stephen Dretler

Patient #3: Popliteal aneurysm

THEN

A 48-year-old female was admitted October 16, 1822 with a popliteal aneurysm. She had fallen on the ice the previous winter and had been confined for three or four days. Initially the leg became swollen and she experienced darting pain from the ankle to the knee. After a few days the swelling had subsided and she went back to work. The following summer, she developed throbbing knee pain and a large mass in the back of her knee, but she was able to walk. By August, her pain had increased and by October she was in such pain that she was unable to stand. On October 16 she was transported to the Massachusetts General Hospital. She was found to have tenderness along the course of the femoral artery. She got warm baths and tincture of opium. Twelve leeches were placed on the area of the aneurysm without result. She underwent venopuncture of 10cc but this caused no improvement.

Two hours after being given six grains of opium, she was taken to surgery. A line was drawn from the inguinal artery to the inner edge of the patella and another line was drawn from the anterior superior spinal process of the ileum to the back of the internal condyle. One and a half inches below the meeting of those two lines, the incision began and extended three inches toward the inguinal artery. Laudanum was used to supplement the opium. The mass was dissected and an aneurysm needle with three silk ligatures was placed under the artery. The aneurysm was 6 inches in length. The ligature was placed under the artery and tightened so that pulsation in the aneurysm ceased. The procedure was done on October 16 and the ligature gently tightened on October 27. Seven days later, December 4, the ligature was removed. It appears that this ligation was above the aneurysm and the aneurysm was left to thrombose. The wound healed without infection and she was discharged.

NOW

The first American surgeon to ligate a popliteal aneurysm was Dr. Wright Post. The concern was that there would be thrombosis and distal embolization. In 1822, it was diagnosed by palpation. In 2011, it is diagnosed by palpation, ultrasonography to assess the femoral and ileac artery for associated fusiform aneurysms, and arteriography, to define the amount of distal runoff. Magnetic resonance imaging (MRI) is currently the most useful method of visualization of the artery and in 2011 is the most accurate method of imaging the arterial wall. It confirms the presence of a mural thrombus and may obviate the need for arteriography in surgical patients, as the status of the adjacent arterial anatomy is clearly defined. In 1949, Dr. Robert Linton of the MGH reported that leaving the aneurysm untreated resulted in limb loss in 77 percent of patients. In the 1930s, lumbar sympathectomy was performed to improve runoff prior to ligation of the aneurysm. In 1912, Hogarth Pringle reported the first saphenous vein graft for revascularization. In 2011, ligation above and below the aneurysm may be performed and replacement of the artery done with saphenous vein, and if the vein is unavailable, with synthetic material (Gore-Tex or Dacron). At the MGH five-year follow-up, 77.2 percent of saphenous vein grafts were patent whereas only 29.5 percent of Dacron prosthesis remained patent. Alternatives include bypass grafting, excision with interposition, and excision with end-to-end anastomosis.

Dr. Ashby Moncure

Patient #4: Urethral stricture

THEN

A 40-year-old male was admitted on October 20, 1821 with a fifteen-year history of urethral stricture. At the age of nine or ten he sustained a perineal injury from a wheel barrel handle. He seemed well for a few years and then around age 25 he noticed his first symptoms of difficulty voiding, accompanied by dysuria, spraying of his stream, and voiding every fifteen to twenty minutes. On October 24, an attempt at dilation by bougies was unsuccessful. Then a catheter was repeatedly passed successfully with dilation of the stricture. He was discharged on November 3, relieved of his symptoms.

NOW

Urethral strictures are as difficult to deal with now as they were then. The urethra is prone to scarring and success is dependent on the length of the stricture and the scarring of the sub mucosal tissues. Diagnosis and treatment have been improved by the use of endoscopic instruments which allow wires to be used to thread through urethral narrowing and gain entrance to the bladder. Instead of blind passage of instruments and the possibility of causing "false passages" and subsequent sepsis, high-pressure balloons are passed over the wire and the stricture dilated. There is controversy regarding endoscopic cutting of the stricture and also whether and how long to leave a catheter in place. If restricture occurs, the distal urethra is marsupialized and "mesh" skin grafts are placed on the exposed urethra and then closed and reconstructed when the grafts are healed. Alternatives for treatment after dilation of distal strictures smaller than two centimeters have included use of buccal mucosal grafts, and primary end-to-end anastomosis. Recently tissue engineering techniques have been used to build an artificial urethra which is a scaffold on which a new urethra is formed. This is new research in 2011 and is not clinically available.

Dr. Stephen Dretler

Patient #5: Anal fistula

THEN

A 40-year-old woman was admitted on October 22, 1821 with two months of rectal pain. On examination of her anus, the opening of a fistulous tract was seen. A probe was passed through the tract and the probe was felt by a finger placed in the rectum. She had bilateral fistulas and a scalpel was used to open the fistulous tracts. Her inflammation was resolving by November 5 and she was defecating without pain. She was discharged on November 15.

NOW

Fistula-in-ano is an ancient disease referred to by Hippocrates and classified by the English surgeon John Arderne in the fourteenth century. Louis XIV was treated for an anal fistula in the seventeenth century. In the late nineteenth and early twentieth centuries, the Lockhart-Mummery grooved director was introduced which allowed for more precise incision of the anal fistula.

Understanding of anal fistula improved in 1900 when David Henry Goodsall introduced Goodsall's Rule, which relates the external opening of the anal fistula to the internal opening. The rule states that the external opening situated behind the transverse anal line will open into the anal canal in the midline posteriorly. An anterior opening is usually associated with a radial tract. Fistulas can be described as anterior or posterior relating to a line drawn in the coronal plane across the anus, the so-called transverse anal line. Anterior fistulas will have a direct track into the anal canal. Posterior fistulas will have a curved track with their internal opening lying in the posterior midline of the anal canal. Exceptions to the rule are anterior fistulas lying more than three centimeters from the anus, which may have a curved track (similar to posterior fistulas) that opens into the posterior midline of the anal canal.

Surgeons of 1821 may have known of the different orientations of the anterior and posterior tracts, but it is likely that some fistulas may have escaped notice. The procedure performed in 1821 was similar to the procedure of 2011, but without anesthesia.

Dr. Paul C. Shellito

Patient #6: Gonorrhea

THEN

The patient was a 48-year-old sailor who entered the hospital on October 25, 1821, with complaints of urethral discharge and incomplete voiding. He gave a history of having a urethral discharge for three months and was found to have incomplete emptying of his bladder. A bougie was passed and a stricture was encountered five inches from his meatus. A larger bougie was then passed and the stricture dilated. It was thought that he had "the Clap" (gonorrhea), but his symptoms resolved after the dilation and he was discharged, relieved.

NOW

If available, antibiotics would have been given prior to any instrumentation. Guide wires would have been placed and the stricture navigated under direct vision to prevent false passages and sepsis. Gonorrhea is a common sexually transmitted infection caused by the bacterium *Neisseria gonorrhea*. The usual symptoms in men are burning with urination and penile discharge. Women, on the other hand, are asymptomatic half the time or have vaginal discharge and pelvic pain. In both men and women, if gonorrhea is left untreated, it may spread locally, causing epididymitis or pelvic inflammatory disease, or throughout the body, affecting joints and valves. Its residual in the urethra is a stricture.

Treatment is commonly with ceftriaxone, as antibiotic resistance has developed to previously used antibiotics.

Dr. Dianne Sacco

Patient #7: Closed long-bone fracture

THEN

On November 10, 1821, a 45-year-old woman was walking down three steps while carrying a pail of water. She fell, striking her leg. She was immediately brought to the MGH with a broken leg. The fracture was unstable and it was noted that both bones in her leg were broken two to three inches above her ankle. The tibia was displaced obliquely, downward and outward. Splints were used and a pillow placed beneath her leg.
By November 12 she had spasms but no pain. Her foot and ankle were swollen and considerably inflamed both above and below the fracture. She received sulfa soda because of slight desiccation just below the fracture. On November 17, seven days after her injury, she had swelling of the calf. By November 27, the bones were in normal position but were still able to be moved at the level of the fracture. By December 4 it was reported the bones were firmly united and the patient had no pain. On December 11 the splints were reapplied. After a hospitalization of 37 days she was able to walk. She was discharged on December 17.

NOW

The cases between September 20, 1821 and May 26, 1822, were some of the first orthopedic cases treated at Massachusetts General Hospital. Each case involved a closed fracture of a long bone resulting from trauma. At that time, the diagnosis and characterization of the injuries was based solely on an accurate history and physical exam. X-rays would not be introduced at MGH until Walter Dodd, an MGH photographer, performed the first hospital-based radiograph in 1907. The resources available and limitations of the times dictated treatments for these long-bone fractures. Laudanum, opium, and sulfa soda were administered for pain control, and bloodletting was performed. Fractures were treated in fracture boxes/cradles (a device constructed of boards and hinges that was used to immobilize fractured extremities in reduced anatomical positions), along with splints and pillows with the goal of creating stability, alignment, and eventual union of the fracture [Figure 1].

Many of the basic principles behind the treatments used nearly 200 years ago for these injuries remain the same today: alignment, stability, and achieving fracture union. However, the advent of anesthesia and sterile techniques, along with improvements in technology has revolutionized the treatment of closed long-bone fractures. In many cases today, long-bone fractures in adults are treated surgically, either with plates and screws, intramedullary rods, or external fixation. Nonetheless, the principles of closed fracture treatment remain an integral part of current orthopedic care.

Patient #32 was a 37-year-old man with a closed fracture of the left humerus, the largest bone in the upper extremity. The patient was given sulfa soda and bloodletting was performed. He spent one month in the hospital and was discharged with a healed fracture. Apart from the prolonged hospitalization, for the most part, modern-day treatment of humeral shaft fractures remains the same. Humeral shaft fractures constitute approximately three percent of all fractures, accounting for approximately 66,000 fractures annually.[1] With more than 90 percent of these fractures going on to union without surgical intervention, non-operative management remains the mainstay for treatment. Surgical indications are limited to open fractures, multiply-injured patients, ipsilateral forearm fractures, and patient intolerance of bracing.[2,3]

Patients with humeral shaft fractures are initially immobilized in a coaptation splint, sling and swathe, Velpeau bandage, or hanging arm cast until initial swelling and pain have subsided [Figures 2a, 2b]. Patients are then transitioned to a functional brace until union of the fracture occurs. A functional brace is an orthosis with an anterior and posterior plastic shell, which provides circumferential hydrostatic compressive forces at the fracture. This stabilizes the arm and leaves the elbow and shoulder free of immobilization to allow motion and avoid stiffness [Figure 3].[2]

In 1977, Dr. Augusto Sarmiento presented his series of 51 patients with humeral shaft fractures treated with functional bracing in *The Journal of Bone & Joint Surgery*, reporting excellent functional outcomes and only one non-union in a patient with a pathologic fracture receiving chemotherapy.[4] These superb results were recently supported by the largest series in print to-date. Dr. Sarmiento followed 620 patients with humeral shaft fractures treated with functional bracing and reported 98 percent and 94 percent union rates for closed and open fractures, respectively.[3] The mainstay of treatment for humeral shaft fractures remains largely unchanged from 200 years ago.

The remaining three cases document patients with traumatic closed tibia and fibula shaft fractures who were treated at MGH in the early nineteenth century. In each case, similar to the man with the humerus fracture, the patients' treatments were dictated by resources available at the time. The patients received sulfa soda and opium, and bloodletting was performed. The fractures were immobilized with cradles, splints, and pillows. Hospitalizations were prolonged with two of three admissions lasting more than one month until union was achieved. Unfortunately, the 56-year-old Scotsman died of an infectious complication of his fracture after only a few weeks.

Apart from the pediatric population, there has been a significant trend toward operative fixation of tibia and fibula shaft fractures. While non-operative treatment of tibial shaft fractures by both casting and functional bracing have been shown to produce good results, certain physician, patient, and fracture characteristics influence the decision to proceed with surgery.[5,6] When treating these fractures a number of factors must be considered: patient compliance with bracing; residual ankle and/or knee stiffness from prolonged immobilization; physician familiarity with casting technique; and certain injury characteristics including soft-tissue injury, initial displacement, orientation of the fracture, comminution, shortening, and ability to maintain alignment (less than five degrees in coronal plane, ten degrees in sagittal plane, shortening of one to two centimeters and malrotation of five degrees considered acceptable).[7]

Furthermore, there is evidence directly comparing conservative management (casting or functional bracing) to operative treatment, specifically intramedullary nails. Tibial shaft fractures with operative treatment results in more rapid time to healing with minimal deformity and less time off of work [Figure 4].[8] Regardless of the treatment method, whether conservative or surgical, good results can be achieved when all aspects of the patient, the injury, and the healthcare environment are taken into account.

Although treatment of closed long-bone fractures at MGH has changed in many ways over the past 200 years, in some aspects it has remained the same. Non-operative management remains the mainstay of treatment for humeral shaft fractures and continues to produce excellent results today. While tibia and fibula shaft fractures can be treated non-operatively in some cases, with improvements in surgical care there has been a significant trend toward operative fixation; both treatment methods yield good results in appropriate patients. Regardless of the long-bone fracture and the treatment provided, the principles underlying the treatment for these injuries remains the same as it was 200 years ago: stability, alignment, and achieving fracture union.

Figures:

Figure 1: Example of a fracture cradle/box used to help align broken limbs while they heal.

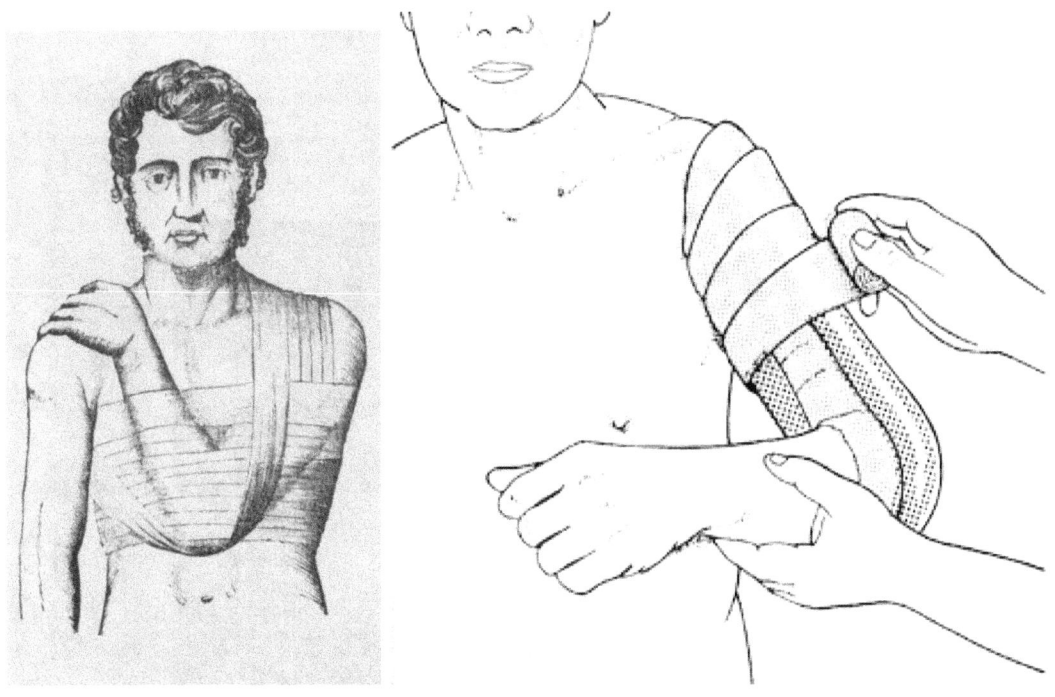

Figure 2: a) Artist depiction of a Velpeau bandage; b) Artist depiction of a humerus coaptation splint.

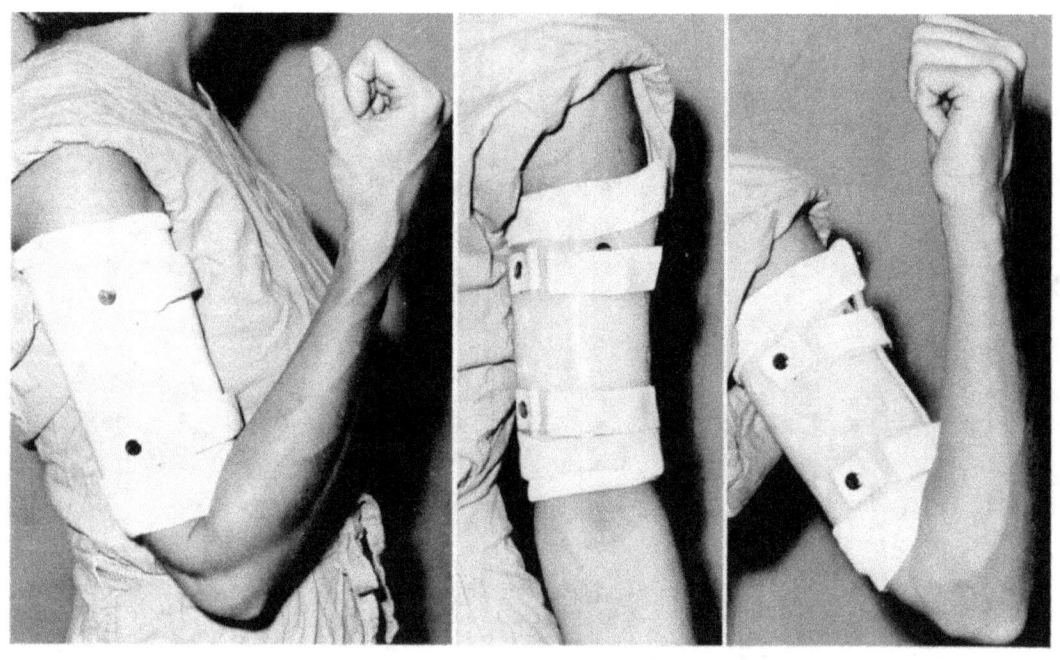

Figure 3: Example of a functional brace for humeral shaft fracture treatment. Taken from Sarmiento's original 1977 article [Sarmiento 1].

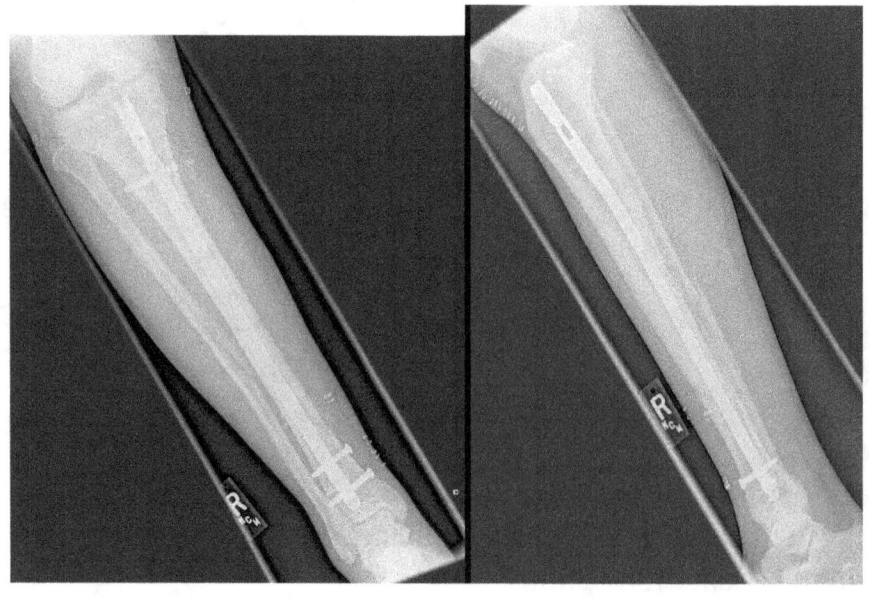

Figure 4: AP and lateral radiographs of a tibia and fibula shaft fracture treated with an intramedullary nail.

References

1. CARROLL, E.A., SCHWEPPE, M., LANGFITT, M., MILLER, A.N. and HALVORSON, J.J., 2012. "Management of humeral shaft fractures." *The Journal of the American Academy of Orthopaedic Surgeons,* **20**(7), pp. 423-433.

2. WALKER, M., PALUMBO, B., BADMAN, B., BROOKS, J., VAN GELDEREN, J. and MIGHELL, M., 2011. "Humeral shaft fractures: a review." *Journal of Shoulder & Elbow Surgery,* **20**(5), pp. 833-844.

3. SARMIENTO, A., ZAGORSKI, J.B., ZYCH, G.A., LATTA, L.L. and CAPPS, C.A., 2000. "Functional bracing for the treatment of fractures of the humeral diaphysis." *Journal of Bone & Joint Surgery - American Volume,* **82**(4), pp. 478-486.

4. SARMIENTO, A., KINMAN, P.B., GALVIN, E.G., SCHMITT, R.H. and PHILLIPS, J.G., 1977. "Functional bracing of fractures of the shaft of the humerus." *Journal of Bone & Joint Surgery - American Volume,* **59**(5), pp. 596-601.

5. NICOLL, E.A., 1964. "FRACTURES OF THE TIBIAL SHAFT. A SURVEY OF 705 CASES." *Journal of Bone & Joint Surgery - British Volume,* **46**, pp. 373-387.

6. SARMIENTO, A., 2007. "A functional below-the-knee brace for tibial fractures: a report on its use in one hundred and thirty-five cases." 1970. *Journal of Bone & Joint Surgery - American Volume,* **89** (Suppl 2 Pt.2), pp. 157-169.

7. LINDSEY, R.W. and BLAIR, S.R., 1996. "Closed Tibial-Shaft Fractures: Which Ones Benefit From Surgical Treatment?" *The Journal of the American Academy of Orthopaedic Surgeons,* **4**(1), pp. 35-43.

8. HOOPER, G.J., KEDDELL, R.G. and PENNY, I.D., 1991. "Conservative management or closed nailing for tibial shaft fractures. A randomised prospective trial." *Journal of Bone & Joint Surgery - British Volume,* **73**(1), pp. 83-85.

Dr. Stephen T. Gardner
Dr. Harry E. Rubash

Patient #8: Hip dislocation

THEN

The patient is a man from Maine admitted to the MGH on December 7, 1821, laboring under dislocation of the left femoral bone. The accident occurred on September 7. He was thrown from a horse and the body of the horse fell across his thigh. It was diagnosed as a dislocation of the thigh and it remained fixed in an oblique position. After ten minutes of forceful attempts to reduce it, the physician told the patient that it had been successfully reduced, but the patient felt no relief and did not have motion in his limb. Three to four hours after the accident the thigh was still unreduced. The physician bled the patient of about twelve ounces and again tried to reduce the dislocation. He again told the patient it was reduced but there remained a tendency for the thigh to assume an oblique position. He was bled another pint and a purgative given. It was thought the neck of the femur was broken. The patient lay on his back for fourteen days and then started to walk on crutches. Two months after the injury it was determined that the femur was dislocated downwards and backwards so that the injured leg was longer than the sound one, and the patient was told that the dislocation had not been reduced. The head of the femur was distinctly felt in the inferior portion of the gluteus muscle. Because of the length of time it had been dislocated, it was decided not to do an operation. The patient was anxious to have a trial at reduction though there was only a small possibility of succeeding. It was thus thought justifiable to attempt a reduction of the bone. On the day of his admission, he was given magnesium sulphate. The next day a warm bath was ordered. He also received a sedative and vomiting was induced. At 3 p.m. he was bled sixteen ounces. No faintness was induced. The operation was immediately started and finally abandoned as hopeless. The patient was discharged on December 10 at his request.

NOW

An addendum to the story is that G.Mason Warren and John Collins Warren commented on this with correspondence and the specimen is in the Warren Museum. We are instructed to see the book on Lowell's Hip Joint, # 347.91 L95 pamphlet, which is kept in the record vault, "Lowell vs Faxon and Hawks."

Dr. James Herndon
Dr. R. Malcolm Smith
Dr. Mark Vrahas

Patient #9: Chronic blepharitis

THEN

An 11-year-old female came to the MGH on December 20, 1821, with salt-rheum (cutaneous eruption) infection involving both eyes. She had had difficulty with it since three years of age. She was given medication and advised to avoid butter and acids. On December 24, she had an oral infusion of 'cinch' twice a day and pills of a form of hydrangea. By January 5, 1822, the inflammation about the eyes was subsiding. A Seton (a stitch passed through a fistula to allow drainage and prevent closure of one end of the tract) was made in the back of the neck. Various anti-inflammatory unguents (semi-paste ointment) were used. Fungous granulations rose up at each orifice. Caustic ointments were applied.

By April 9, 1822, she had no inflammation about the eyes except for the edge of the eyelids and anti-inflammatory ointment was placed on the lids using a "hair pencil." By June 27, her eyes were better and she was discharged with instructions to apply a poultice and use unguent of ophthalmic, and by September 11, she was discharged.

NOW

This is a case of a child with a chronic blepharitis. Based on the clinical description there are eczematous lesions on her eyelids and fungal granulations (unclear if they are located at the edges of the fistula or in the eyelid lesions). A differential diagnosis for this clinical scenario would include atopic dermatitis, recurrent herpetic blepharitis, as well as focal granulomas or dermatitis secondary to blastomycosis, sporotrichosis, rhinosporidiosis, cryptococcosis, leishmaniasis, or ophthalmomyasis. Tuberculous blepharitis could be another possibility. In general, work-up is not necessary for typical cases of blepharitis. Demonstration of HSV can be done with viral cultures or antigen/DNA detection methods. Serologic testing can be done but is not useful during the acute episode and only helpful when it is negative. KOH prep can be useful in cases of fungal infections and an acid fast stain is the appropriate test for detection of mycobacteria. A simple gram stain and culture can sometimes be helpful.

In the case of atopic dermatitis, topical antihistamines and corticosteroid creams can alleviate symptoms and in the case of herpetic blepharitis, oral antivirals (e.g., acyclovir) can speed the resolution of signs and symptoms. In the case of fungal etiologies, topical antifungal agents can be employed.

The physicians of 1821 had no antibiotics at their disposal and germ theory was in its infancy. Cultures were not possible. Consistent with the use of leeches or bloodletting, the creation of an iatrogenic fistula tract was presumably intended to draw the illness out from the patient. Needless to say, admission for blepharitis today is unheard of. It is ironic, however, that the most effective treatment for blepharitis remains the least technologic and most available to the physicians of 200 years ago: warm compresses and lid hygiene.

Dr. James Chodosh
Dr. Matthew F. Gardiner
Dr. Joan W. Miller
Dr. Sotiria Palioura
Dr. Athanasios Papakostas

Patient #10: Inflamed testicle

THEN

A 31-year-old male was admitted to MGH on January 4, 1822 with a "hernia humoralis" (inflamed testicle). The testicle was inflamed and swollen and had been since September, 1820. He has needed to use a suspensory bandage since that time. He was treated with a form of hydrangea (an unguent) twice a day and warm baths twice a day. He also was to rub his thighs with ointment twice a day. By January 12, eight days after starting treatment, he had a sore mouth and pills and ointments were omitted. By January 21, the testicle swelling was somewhat diminished. Finally, a seton was proposed and refused. He was discharged on February 15 at his own request, "considerably relieved." There was no recorded follow-up.

NOW

A 31-year-old male was admitted to MGH on January 4, 1822, with an inflamed and swollen hemiscrotum since September, 1820. Given the length of time of the symptoms, it is unlikely that the gentleman had an acute condition such as epididymitis/orchitis or torsion of the spermatic cord. He did not seek care for almost sixteen months, and he seemed to get by with a suspensory bandage. Viral orchitis, such as mumps orchitis, would have subsided earlier, usually leaving the patient with an atrophic testis. Other nonbacterial orchitides—fungal, parasitic, or rickettsial—would be associated with continuous, worsening illness in a toxic, febrile patient. It is also possible that he had a noninfectious orchitis, either traumatic or autoimmune. If he truly had a "hernia humoralis," he may have had a hydrocele or an indirect inguinal hernia with a hydrocele. It is noteworthy that, seventeen days after initiation of treatment with an unguent, warm baths, and thigh rubs with an ointment, the swelling was "somewhat diminished." Hydroceles can vary in size over time. He spent 42 days in the hospital, and he was discharged at his own request, "considerably relieved."

Now, the patient would be managed with a physical exam to rule out the possibilities mentioned above, and, perhaps, with an ultrasound of the scrotum to further assess the problem. Testicular torsion in young men is diagnosed with the ultrasound and Doppler imaging to assess testis blood flow. Torsion of the appendix epididymis or the appendix testis usually resolves spontaneously. If the epididymitis is infectious in nature, appropriate antibiotics, after obtaining the necessary cultures, are administered, usually as an outpatient, unless the patient is toxic and febrile. Young men, suspected to harbor a sexually transmitted disease, are cultured with a urethral swab for the most common organisms, namely Neisseria gonorrhoea and Chlamydia trachomatis. Mycobacterial infections may also cause epididymitis/orchitis. Chronic orchits/orchalgia is treated with supportive measures, such as suspensory bandage, anti-inflammatory agents, analgesics, and, perhaps, nerve blocks. Surgery for hernia/hydrocoele is usually only indicated if the condition interferes with the patient's activities of daily living. Epididymectomy/orchiectomy is only used as a last ditch measure when all else has failed.

Dr. Adam Feldman

Patient #11: Vertebral trauma

THEN

The patient was a 27-year-old Irishman who was brought to the hospital on January 4, 1822, with a spinal injury. He was shoveling under a roadway and the road gave way and the patient fell twelve to fourteen feet, burying him to his shoulders. Three or four hours later he was brought to the hospital with much pain referred to the small of his back and was unable to take a full respiration. He was bled for 20 ounces of blood which immediately relieved his respiratory distress. A fomentation (warm, moist medicinal compress) was applied to his back and a cathartic was given.

Two days later examination showed that one of his vertebrae was displaced. Four days later he was restless and had difficulty passing urine, voiding frequently and in small amounts. In addition to flaxseed tea and Tart Pot Soda, six leeches were applied to his spine. By January 15 he felt much better but turned in bed with difficulty.

By January 30 he was able to walk in the ward without pain. On February 5 he had back weakness but was discharged well on February 18.

NOW

The man in this scenario received a large, compressive force to his spine. The bulk of the force was likely directed to the shoulders and upper thoracic spine as one typically would bend forward and tuck the head when debris begins to fall from above. This compression/flexion mechanism would lead most commonly to injuries in the thoracolumbar spine. This mechanism is similar to what one would see in mining injuries during a "cave in." Indeed, thoracolumbar fractures have been reported to represent 66 percent of all spine fractures seen in miners.[1] The pain and deformity noted in "the small of his back" when combined with the mechanism of injury is consistent with "burst" fracture of the lumbar spine. The most common location for such an injury is the junction between the lower thoracic spine and the upper lumbar spine. Given the reference to the small of his back it is likely that the fracture occurred at the first or second lumbar spine. The man had a "displaced" vertebrae and this likely represented a gibbus deformity which is taken from the Latin *gibbosus,* or humpbacked. If this patient were injured today an ambulance would likely have been called.

Given the mechanism of injury and his complaints of pain in his back, he would have been placed in a hard cervical collar and secured to a rigid board. Intravenous access would have been obtained and he would have received crystalloid. Depending on the level of pain he would be given some type of pain medication, with morphine sulfate being the most common. Once he arrived in the emergency room, and life threatening injuries had been addressed or ruled out, a spine consultation would be obtained. A thorough neurologic examination of his cranial nerves, trunk, and extremities, including a rectal examination, would be performed. Particular attention would be paid to the contour and sensitivity of the spinous processes of the spine. It is likely that the thoracolumbar spine would be sensitive. In severe cases ridge that is exquisitely tender may be noted. This is often indicative of an injury to the posterior elements of the spine and when combined with an anterior spinal injury is usually unstable. Given the rapidity with which the man recovered from his injury, I doubt that he had this severe of an injury. I suspect that his injuries were confined to the vertebral body alone. Computed tomography (CT) is readily available and today this patient would likely have a CT of his spine rather than starting with plain radiographs. Depending on the results of the CT he may also have an MRI to evaluate the ligaments in the posterior aspect of his spine. Since the man's injury predated the development of plain radiographs by 73 years, the treating physicians did not have their benefit.

The patient likely had a burst fracture and was neurologically intact upon presentation. However, after a week in the hospital he began having trouble passing urine. While this may be only indirectly related to his fracture, one has to be concerned about the development of cauda equina syndrome secondary to bony compression of the lumbo-sacral nerve roots. He likely had a burst fracture involving the vertebral body with retropulsion of bone into the spinal canal. The most sensitive symptom of cauda equina syndrome is urinary retention. If this were to happen today he would be questioned about numbness in his perineum and specifically about numbness in his rectum. This would be followed by physical examination of the area. He would likely have an MRI at this point as well to evaluate the spinal canal.

The treatment of burst fractures has evolved over time. Most patients in the 20th century would be treated with reduction in hyperextension followed by casting.[1] In 1983, a seminal article in spine surgery was written by Francis Denis touting the importance of the "middle column." He found a high rate of neurologic injury, when the middle column was compromised with a fracture.[2] The middle column includes the posterior half of the vertebral body and disc as well as the posterior longitudinal ligament. Burst fractures by definition involve the middle column and so it would be predicted by Denis that many patients with this injury pattern would develop permanent neurologic sequelae.

Denis' article changed the way burst fractures were treated from one which relied on reduction in hyperextension followed by casting to a more surgically based practice of spinal fusion. The tide began to change again when an article published by Wood et al. compared operative vs. non-operative treatment of thoracolumbar burst fractures and found that there did not seem to be an advantage for surgery.[3] Furthermore, they did not note neurologic decline with non-operative treatment. The non-operative management, in the study by Wood et al., included a brace. The patient was not treated with surgery or a brace and he recovered quite nicely without either. Interestingly, there has been a recent study that compares non-operative treatment with and without a brace for burst fractures. This study found that bracing did not seem to add any benefit when compared to not bracing.[4]

In this instance, the patient was walking without pain within one month, and although he was described as having back weakness at discharge, he presumably walked out of the hospital and went on with his life two months after his injury. This patient might have been treated operatively if he were seen ten to twenty years ago at MGH. However, there is mounting evidence that not only is surgery unnecessary, as is bracing or casting in many cases. So it seems that the treatment rendered in 1822 for this fracture comes close to how it would now be treated at Massachusetts General Hospital—now sans the leeches.

1. Nicoll, E.A., *Fractures of the dorso-lumbar spine.* J Bone Joint Surg Br, 1949. **31B**(3): p. 376-94.
2. Denis, F., *The three column spine and its significance in the classification of acute thoracolumbar spinal injuries.* Spine (Phila Pa 1976), 1983. **8**(8): p. 817-31.
3. Wood, K., et al., *Operative compared with nonoperative treatment of a thoracolumbar burst fracture without neurological deficit. A prospective, randomized study.* J Bone Joint Surg Am, 2003. **85-A**(5): p. 773-81.
4. Bailey, C.S., et al., *Comparison of thoracolumbosacral orthosis and no orthosis for the treatment of thoracolumbar burst fractures: interim analysis of a multicenter randomized clinical equivalence trial.* J Neurosurg Spine, 2009. **11**(3): p. 295-303.

Dr. Joseph Schwab

Patient #12: Compound femoral fracture

THEN

On January 4, 1822, a 22-year-old male was admitted with a fracture of the thigh and a compound fracture of the tibia. On January 6, a fracture box was applied with adhesive straps and a pint of blood was withdrawn. He was stimulated to have three bowel evacuations. By January 9, the wound was not inflamed and it was allowed to heal by secondary intention. Pledgets were placed on the open wound and the wound was dressed daily. The patient had no significant symptoms. On January 22, it was determined that the bones were perfectly united. By February 1 the leg wound had healed. On February 5 the fracture box was removed from his limb and on February 18 he was discharged.

NOW

Change is inevitable in the practice of medicine, especially in the treatment of open fractures. This is evidenced by the preceding orthopedic case and others that took place at the Massachusetts General Hospital between January and November of 1822.

If not treated appropriately this type of injury can cause significant problems, the most common being superficial or deep infection involving the skin, deep tissues, or bone, and potentially systemic infection and sepsis. Current classifications exist to describe the severity of these injuries: Type 1 is a small, clean wound; Type 2 is a slightly larger wound without extensive soft-tissue damage; and Type 3 is a large wound with significant soft-tissue loss, bony exposure or comminution, and possible neurovascular injury.[1]

Since the fracture site is now exposed to the environment, a common complication of these types of fractures is infection. Management of these injuries, regardless of severity, has changed significantly over the past 200 years. In the nineteenth century, clinicians had to rely solely on a thorough history, physical examination, and clinical acumen to arrive at a diagnosis. Modern medicine equips current physicians with sophisticated diagnostic tools and imaging to aid in diagnosis, a luxury that was not present in 1822. Treatment of orthopedic ailments, including open fractures, consisted of splinting and conservative measures; operative intervention was utilized for progressive or severe cases. Pain management for operative procedures consisted of laudanum, a tincture of opium, which is an herbal preparation containing approximately ten percent powdered opium, or the equivalent of one percent morphine. Other hallucinogens were also used. The concept of anesthesia was not realized until October 16, 1846, when the first public demonstration of ether anesthesia took place at the Massachusetts General Hospital, a seminal event in the history of surgery and medicine.

This patient sustained a blunt trauma to his lower extremity and suffered a closed femur fracture and an open tibia fracture. Initial treatment consisted of a fracture box, which is akin to modern day splinting techniques, as well as pledgets (small absorbent cotton strips) placed in his open fracture wound. He spent approximately six weeks in the hospital and was discharged with what was described as a united fracture. Current evidence-based practice supports a more aggressive approach to open fractures. Most of this research was pioneered by Drs. Raymond Gustilo and John Anderson in the early 1970s. In keeping with the theme of an ever-evolving practice and the dynamic change in medical treatments, our current concepts in open fracture treatment have changed significantly in just the past thirty years since the tenets introduced by Gustilo.

Gustilo et al. stated that open fractures require emergency treatment including adequate debridement and irrigation of the wound, but several questions were left unanswered. These questions involved antibiotic use, whether the overlying wounds should be closed immediately, and whether the underlying fracture should be fixed immediately. These questions would hopefully address the problems of infection with open fractures. Gustilo retrospectively analyzed 673 open fractures treated at his institution, which were all managed with adequate irrigation and debridement. Primary closure with internal fixation was used whenever possible in conjunction with routine antibiotic administration.

The results showed that primary internal fixation increased the incidence of infection, that prophylactic antibiotics are essential in treatment, and that a higher severity of injury resulted in a higher infection rate. A prospective study was also performed in which 352 patients were managed with thorough irrigation and debridement, primary closure for Type 1 and Type 2 fractures, no primary internal fixation except for the presence of associated vascular injuries, cultures of all wounds, and antibiotics before surgery and three days post-operatively. He found a significantly decreased level of infection with this methodology.[2]

This research helped shape our current understanding of and methods for dealing with open fractures. In this case, no irrigation or debridement was carried out and no fixation was completed. Antibiotics were not yet utilized in practice until the discovery of penicillin in 1928, which is attributed to Scottish scientist and Nobel Laureate Alexander Fleming. Today, this patient would be assessed, irrigated, and primarily fixed if the soft tissues warranted. The patient had a splint applied to help align the fracture and the open wounds were treated with serial dressing changes. A major tenet in the treatment of open fractures today is that stabilization, in conjunction with adequate soft-tissue coverage, is pursued at a very early stage in treatment. This patient's tissues were allowed to granulate via secondary intention, rather than by surgical closure. Today, he would be managed with thorough irrigation and debridement of devitalized tissues, stabilization of the fracture, and primary closure versus skin grafting/flap coverage to ensure a healthy soft-tissue envelope for healing.

The second case of compound fracture (Patient #14) involves a young man who also sustained blunt trauma to the lower extremity when an axle from his cart became dislodged. He was diagnosed with an open tibia fracture. His open wounds were treated with alcohol to help prevent infection. His wound eventually became gangrenous and a small portion of exposed bone needed to be removed. Wine and laudanum were given for pain control. His gangrene and infection continued to progress and he eventually required amputation. This patient subsequently died two weeks after admission to the hospital. This is an unfortunate complication of open fractures, which aggressive early treatment currently attempts to prevent. This patient likely had significant infection and possibly an unrecognized vascular injury and may have had a different result had these issues been addressed early.

The question of timing of debridement has become a common theme in the management of these fractures as well. In 1881, Karl von Reyher, a Prussian Minister of War, demonstrated decreased mortality rates with early debridement of open injuries during the Franco-Prussian War.[3] This was followed in 1898 by P. Friedrich's experimental tissue contamination studies of guinea pig soft-tissue wounds, which established the time interval for the effectiveness of operative debridement to be approximately six hours.[4] This was challenged by Harley et al. with research stating that time was not a significant factor in preventing non-union or infection.[5] In 2012, Schenker et al. stated that data did not indicate an association between delayed debridement and higher infection rates and noted that the historical six-hour rule proposed by Friedrich has little support in the available literature.[6]

With this being said, the basic principles of treatment of open fractures still apply: early and adequate debridement and wound care, early coverage and closure, appropriate skeletal fixation, and early administration of antibiotics. If these principles had been applied as they are today, this patient's limb may have been salvaged and he ultimately may not have perished.

The third and fourth cases of compound fracture (Patients #19 and #53 respectively) further demonstrate the conservative treatment measures employed by physicians in the early nineteenth century. Patient #19 was a 37-year-old male who had an open fracture of the tibia and fibula. He was splinted (as the leg was not "set properly" in the case description) and eventually left the hospital without any debridement, albeit two months after admission. Patient #53 is a 26-year-old male who was struck in the shoulder by a rock and sustained a proximal humerus fracture with a fistulized infection or abscess. The infection was treated with turpentine and rubbing oil. He eventually underwent local debridement and subsequent operative debridement and was discharged from the hospital after four months of treatment. These cases further support the notion of aggressive early treatment, not only to help to decrease the morbidity and mortality associated with open fractures, but also to decrease the length of stay for patients, which secondarily helps decrease complications and improves the patient's quality of life.

A concepts review conducted at our institution by Drs. Okike and Bhattacharyya[7] helps define our current thoughts regarding open fractures. The concepts are:

- To reduce the risk of infection, antibiotics should be administered to a patient with an open fracture as soon as possible.
- A patient with an open fracture should be taken to the operating room on an urgent basis with the stability of the patient, the preparation of the operating room, and the availability of appropriate assistance taken into account.
- Questions remain regarding the optimal solution and method of delivery for irrigation of open fracture wounds.
- Early closure of adequately debrided wounds is safe and can improve outcomes.
- Adjunctive therapies, such as the early application of bone grafts and rhBMP-2, may improve the healing of open fractures.

Treatment of open fractures at MGH has changed significantly over the past 200 years. Early operative management remains the mainstay of treatment today, whereas conservative measures were largely employed in the 1800s. This was primarily an institution of the times, rather than any negligence on the part of the treating physicians. The discovery of anesthesia and antibiotic use in medicine has truly revolutionized practice in all fields, including orthopedic trauma and the treatment of open long-bone fractures. Evidence-based medicine drives our current practice of treatment. Only time will tell if we have the proper answers for treating these conditions.

References
1. Gustilo RB, Mendoza RM, Williams DN. Problems in the management of type III (severe) open fractures: A new classification of type III open fractures. J Trauma. 1984; 24:742–

2. Gustilo RB, Anderson JT. Prevention of infection in the treatment of one thousand and twenty-five open fractures of long bones: Retrospective and prospective analyses. J Bone Joint Surg Am. 1976; 58:453–8

3. Wangensteen OH, Wangensteen SD. Carl Reyher (1846-1890), great Russian military surgeon: His demonstration of the role of debridement in gunshot wounds and fractures. Surgery.1973 Nov; 74(5):641-9.

4. Friedrich PL. Die aseptische Versorgung frischer Wundern. <u>Arch Klin Chir</u>. 1898; 57:288-310.

5. Harley BJ, Beaupre LA, Jones CA, Dulai SK, Weber DW. The effect of time to definitive treatment on the rate of nonunion and infection in open fractures. J Orthop Trauma. 2002 Aug; 16(7):484-90.

6. Schenker ML, Yannascoli S, Baldwin KD, Ahn J, Mehta S. Does Timing to Operative Debridement Affect Infectious Complications in Open Long-Bone Fractures? A Systematic Review. J Bone Joint Surg Am. June 2012; 94: 1057-1064.

7. Okike K, Bhattacharyya T. Trends in the management of open fractures. A critical analysis. J Bone Joint Surg Am. 2006 Dec; 88(12):2739-48.

Dr. Jeffrey R. Jaglowski
Dr. Harry E. Rubash

Patient #13: Urethral stricture

THEN

A 52-year-old man was admitted January 9, 1822. He had been sick for about two years and had lost his desire to pass urine, had severe dribbling and passed only a small quantity of urine at a time. He also had a purulent and bloody urethral discharge. A stricture was found four to five inches from the urethral meatus. A small bougie passed daily did not dilate the stricture. On January 29, a caustic bougie was used to enter his urethra and caused great pain. He was repeatedly dilated with a caustic bougie passed over a dilating bougie. Urine with purulent matter was passed and on March 23, a bougie of considerable size was pushed into the bladder.

He was asked to self-catheterize daily. His prostate was enlarged. His urine was constantly dribbling and when drained by catheter appeared bloody and purulent. Despite injections of laudanum (tincture of opium), he had great pain and straining. The patient died on June 10 almost five months after his admission.

NOW (See Patient #4)

Patient #14: Compound tibial fracture

THEN

A black man was brought into the hospital on January 26, 1822 with a compound fracture of the right leg. Three days earlier a wheel had come off the axel of his cart and struck him in the leg. The night before he was admitted he had been delirious and medicated with 50 drops of laudanum. The leg was kept wet with alcohol. The morning of admission the leg and foot were very swollen. By January 30, the swelling was a little diminished. Small areas around the wound were gangrenous. By January 30, a portion of the fractured tibia was exposed and about one and a half inches of bone was sawed off and the leg dressed again. He was allotted two glasses of wine per day. By February 5 the wound looked gangrenous and preparations were made for amputation. Instruments used were: two knives, a sixteen-inch-long and ten-inch-wide cutting saw, a tourniquet, ligatures of twisted silk thread, vessels of warm water, and a piece of chalk.

The patient was purged with rhubarb and given a strong dose of laudanum. He was placed on the table with each leg and arm held by an assistant. A tourniquet was placed mid-thigh and amputation was carried out on February 5. Ligatures of thread were used. An adhesive was placed and a dressing of strong linen and cotton were placed.

The patient died on February 11, 1822.

NOW (See Patient #12)

Patient #15: Venereal disease

THEN

The patient was a 22-year-old man from Portland, Maine who contracted gonorrhea and had a discharge for three to four weeks. He was admitted on February 4, 1822. A small ulcer appeared on the glans penis and enlarged over the next twelve days. An operation for phimosis was performed. His urine was seen to pass through the side of his penis. Ulceration extended from halfway down the length of the penis to the glans. On February 28, the wound looked well and the ulcer was almost healed. A silver nitrate solution was regularly applied to the ulcer. He was discharged on April 2, after nearly two months of hospitalization.

NOW

Dermatology cases from the Massachusetts General Hospital from the 1820s reveal a host of maladies, from syphilis to herpes to eczema. An overwhelming theme in the cases from that time is the genital ulcer. Historically, dermatology and venereology were linked as a single field, as can be seen in early dermatologic texts, such as Josef Jadassohn's Handbook of Skin and Venereal Disease.

Over time, dermatologic care has shifted from a focus on the infectious to the neoplastic. According to records from McCall Anderson[1] in Scotland in 1887, at least 50 percent of 11,000 recorded cases were infections or infestations, in contrast to the modern Scottish Dermatology Clinic where close to 50 percent of visits were for skin cancer.[2] Our own experience at MGH mirrors this trend.

The realm in which we treat patients has also shifted. While at MGH we have a growing inpatient dermatology consult service, which sees a variety of inpatients with leukemia, infection, drug eruptions, and neutrophilic dermatoses, much of present day dermatology care is done on an outpatient basis. Weekly wound care clinics for ulcer management have replaced the two months of hospitalization experienced by the patients described in these early MGH records.

---

[1] McCall Anderson T. *A Treatise on Diseases of the Skin*. London: Charles Griffin and Co, 1887.
[2] Hunter, J.A.A. "Turning points in Dermatology during the 20th century." *British Journal of Dermatology*, 2000; 143:30-40

Some treatments used then are still used today. For example, silver nitrate, used for ulcers (in this case and see also Patient #72), is still in common use in the form of silver treatment and silver impregnated dressings. A recent Cochrane review and meta-analysis found that silver impregnated dressings improve reduction in wound size, but there is less data about long-term effects on wound healing.[3]

Mercury, which was proffered as a treatment (see Patients # 31, 72, and 78), has been in dermatologic news of late as a potential toxic additive to skin creams. According to the Environmental Protection Agency, several states have investigated cases of mercury poisoning due to the use of unlabeled skin lightening creams imported from Mexico and the Dominican Republic.[4] A recent U.S. Centers for Disease Control report warns that they have identified cases in both California and Virginia where patients with high levels of mercury were using creams containing thousands of times the level allowed by the U.S. Food and Drug Administration.[5] We wonder about the long-term effects that mercury pills, ingested by our patients in the 1820s for their skin complaints, might have had on their mortality.

Dr. Esther Freeman

Patient #16: Frostbite of toes

THEN

An 18-year-old female was admitted on February 4, 1822, with frozen gangrenous feet. She was an epileptic and had been afflicted since the age of four. Her fits would last 12-24 hours, sometimes with days between fits. She was now like an "idiot." The night she got frostbitten she was sleeping in a room with a fire burning all night. She may have gone outside because her feet and hands were purple.

Her hands recovered in three to four days but her feet worsened. On February 5, an oatmeal poultice was used. On February 10 the toes on both

---

[3] Carter, MJ, Tingley-Kelley K, Warriner RA. "Silver treatments and silver-impregnated dressings for the healing of leg wounds and ulcers: a systematic review and meta-analysis." JAAD, 2010 Oct;63(4):668-79.
[4] U.S. Environmental Protection Agency: http://www.epa.gov/hg/consumer.htm#creams Accessed 2/2/12.
[5] U.S. Centers for Disease Control and Prevention (CDC). *MMWR Morbidity and Mortality Weekly Report*, 2012 Jan 20; 61:33-6.

feet were removed and her feet were dressed with dry lint. The bones of the metatarsals were covered with good granulations and projected about one third of their length through the granulations. The ends of the three intermediate bones were just visible By April 12 her feet were gradually healing. However, she had fits which caused her feet to bleed and retarded healing. On July 3 she was discharged, feet "cured."

NOW

Most cases of frostbite are encountered in soldiers, in persons who work outdoors in the cold, in homeless people, in athletes engaging in sports with seasons extending into the cold months of the year, and in winter outdoor enthusiasts, such as skiers. In some patients seen with frostbite today, alcohol and illicit drug use rather than mental disorders characterize individuals who present with cold-weather-related injury; these individuals have impaired judgment and they may not sense that they are in danger.

The initial treatment for any cold-weather-related injury involves removing the patient from the precipitating cold environment to prevent further heat loss. It is essential to remove all wet clothing and constricting clothing (such as socks, boots, and gloves), and replace with dry clothing. It is important to be sure that the patient is not suffering from hypothermia, a potentially life-threatening condition. In extreme cases of systemic hypothermia, patients may be warmed by peritoneal lavage or veno-venous bypass.

Prior to transport to a health care facility, if possible, the affected area would be wrapped in a dry sterile bandage or a clean blanket to prevent further trauma. Cotton may be placed between the toes or fingers, if affected, to prevent any potential damaging effects of rubbing against one another.

On arrival in the ER, life-threatening issues would be addressed first. Fluid resuscitation, especially in persons with mountain frostbite, enhances blood flow and tissue perfusion. The affected body part would be rapidly rewarmed, avoiding further trauma. The most effective treatment measure for frostbite is rapid rewarming. This is accomplished by immersing the affected area into a circulating tub of warmed water that is 40-42 °C (104-108 °F) for 20 to 40 minutes or until thawing is complete.

Dr. Michael Watkins

Patient #17: Venereal disease

THEN

A 29-year-old male patient was admitted July 16, 1822. In the winter of 1813 he had been sick with "lung fever" and ague (malaria). He was sick for about two months and then well for about two years when fever recurred. He was in Plattsburg Hospital for five months. His health had been poor since that time. He was occasionally plagued by cutaneous eruptions and red color around his arms and legs. He had severe itching, especially when approaching fire. He had contracted gonorrhea nine years earlier, which disappeared in two to three weeks without apparent complication. Four years ago he had venereal disease first seen as a black spot on his glans penis which gradually enlarged to the size of a thumbnail and then ulcerated, healing in two to three weeks, followed by swelling in the groin. In July, 1822, he had fever and ague and was getting over that, when an eruption occurred on arms and legs and quite covered his forehead and face. His legs became swollen. In October, 1822, his eyes were inflamed and the eruptions on his head and face disappeared. He had several "nodes" on his tibia. His eyes remained inflamed and that side of his face was very painful and secretions from the nostril on the same side as the inflamed eye had increased.

The last eruption did not itch upon approaching fire but was very painful if exposed to the cold. He had several small ulcers on his legs and was unable to open his eyes. He had 'darting' pain in his left eye and on that side of the face. He was treated with various topical ointments including belladonna. On February 19, his pain was reduced. On March 10 his eruption disappeared. On March 26 he still had some eye inflammation and was treated with a topical sulfur eye wash. On April 2 his eyes were better and he was discharged on April 5, "cured."

NOW (See Patient #15)

Patient #18: Cellulitis of toe

THEN

A 23-year-old woman was admitted on February 28, 1822. Two or three years earlier she had had ulceration on the side of her great toenail as a consequence of wearing tight shoes. Some of the nail became loose and was taken out. A new nail started to grow and the ulceration restarted. She had pain and inflammation from toe to hip. Pain medication and a simple poultice were given and she was ordered on strict rest. By March 5 she had very little pain. On March 16 the nail was cut out and dressed with lint. She was kept on rest. The healing process and regrowth of hard substance (nail) followed the groove of the old nail and she was discharged on June 12.

NOW

A 23-year-old patient with a toe ulcer from repeated trauma had neglected the wound and presented with cellulitus surrounding the toe and spreading to the foot and leg. A serum glucose was drawn and plain films of the toe were ordered to evaluate for osteomyelitis. The patient was a diabetic and appropriate treatment was started. She was admitted for intravenous antibiotics, surgical debridement of the ulcer, and removal of the nail.

Dr. George Velmahos

Patient #19: Open oblique leg fracture

THEN

A 37-year-old male patient was brought to the hospital on March 8, 1822. Both bones of the left leg were fractured obliquely and the leg had not been set properly. In various places on the leg there was skin breakdown. He had delirium tremens since March 5. A fracture box was used. On March 17 the leg was examined and found to be in good position but without union. The fracture box was removed and splints were applied. By April 2 the bones were nearly united, with just a little motion.

On April 18 the leg was noted not to have united properly. By April 25 the bones appeared united and the splints were left off. The leg was swollen but strong. Patient was discharged "cured" on May 24.

NOW (See Patient #12)

Patient #20: Stab wound of leg

THEN

A 16-year-old male was admitted to "the house" on May 24. He had received a wound in the back of his leg caused by shears being thrown with such violence that one blade penetrated through his leg. By June 17 he was healed and discharged "cured."

NOW

This patient would be evaluated by physical exam and possibly CT angiogram to ensure no nerve or major vascular injury was incurred. He would receive a tetanus toxoid immunization and potentially had the wound washed out in the emergency room. A single dose of antibiotics may have been given, and if no vascular injury was suspected, this patient might have been discharged either from the emergency department or after an overnight stay in the hospital.

Surgical Service

Patient #21: Penetrating injury of skull

THEN

A 28-year-old man entered "the house" on March 13, 1822. He had a wound in the forehead above the frontal sinus, the consequence of discharging a "fowling piece" which burst. The gunstock was splintered and one of the screws struck him in the forehead. He was senseless for a few minutes. The date of the incident was December 6, 1821, almost two months previously. The area where he had been struck by the screw had not healed and it kept up a discharge. On the outside the wound was small. A probe could be put into it. The external plate of bone had been broken and he had not been well since the accident. He attempted to work but experienced too much pain in the head. He was treated with topical nitrates, silver nitrate, distilled water, a syringe to the wound twice a day, and dressed twice daily with an ointment.

On March 16, fungating granulomatous tissue projected from the upper point of the wound. Caustic potash was applied but did not relieve his pain. March 24, eleven days after his admission, the wound appeared healed and he was discharged on March 27. On June 2 he was seen at the hospital again because he sprained his knee, but his head was healed.

NOW

At the time of the injury, Emergency Medical Services would be called, and the patient would be transported by ambulance to the nearest Level 1 Trauma Center. Because he had a loss of consciousness, and because there might be concern about a major penetrating traumatic brain injury that might lead to rapid deterioration, he might be sedated and intubated at the scene prior to transport, using protocols designed to control intracranial pressure. A team would be waiting for his arrival to the Trauma Bay consisting of specialists in Emergency Medicine, Trauma Surgery, Radiology, and Neurosurgery. After immediate attention to any necessary resuscitation and intravenous access for medications or blood transfusions, an initial rapid but thorough step-by-step full evaluation would be done by protocol to assess general status as well as specific focal injuries. The patient next would be transported to another area within the Emergency Department for imaging - in this case, CT scan, possibly including CT angiogram and/or venogram - which can assess both the bony and intracranial damage wrought by the firearm projectile. Since this was an open depressed skull fracture involving the frontal sinus, a surgical team of appropriate specialists would be assembled. Urgent surgery would be undertaken to debride the wound, exenterate the frontal sinus, repair any posterior wall and/or dural injuries, and reconstruct the bony and soft tissue layers. Cultures would be obtained and broad-spectrum antibiotics would be administered. If the missiles caused widespread contamination or fragmentation precluding primary repair, a second procedure might be planned to reconstruct the anterior cranial vault at a later date. The patient would be observed post-operatively in a surgical or neuroscience critical care unit by specially trained nurses and intensivists, and a variety of specialists would monitor and take steps to prevent infection, promote wound healing, and address any sequellae of injury to the frontal lobe, including the prevention of early seizures. After discharge, the patient might receive outpatient rehabilitative therapies, and would be followed at regular intervals to monitor for delayed complications and to maximize functional and cosmetic outcome.

Dr. Ann-Christine Duhalme

Patient #22: Urethral stricture

THEN

A male, age 54 and a native of Dedham, came to the hospital on March 23, 1882. About ten years previously, he had contracted gonorrhea. He used injections and calomel for two to three weeks and the gonorrhea ceased. He had some swelling of the groin and neck, as well as difficulty and pain making water and frequent inclination to void, often times with violent straining. A few months after the gonorrhea had ceased, he had a small sore on the septum of his nose which was not painful but in a short time, caused an opening through the septum. Within six months, this opening had enlarged and very often, drained a bloody scab from the nose in the morning. For the past three years, he had pain in the legs, particularly in the night. He also had incontinence of urine with discharge of matter from his urethra when voiding his water. Three years earlier, he had a fever and ague. His urine was highly colored with sediments. The previous summer, he had trouble with an eruption resembling impetigo. On March 23, he was found to have a stricture, three quarters of an inch from the external orifice of the urethra, formed apparently on the opposite side of the urethra.

With some difficulty, a small bougie was passed and another nearly four inches beyond the obstruction. The proximal stricture was impervious to the small bougie. The bougie was applied again in the afternoon. A larger bougie was passed along the stricture but would not enter the bladder. On April 9, a small catheter was introduced. This was done every day, increasing the size of the catheter. He was discharged "cured."

NOW (See Patient #4)

Patient #23: Corneal abrasion

THEN

The patient, age 25, entered "the house" on March 23, 1822. It had been five years since he had infections of the eye, caused, he thought, by working in verdigris. He came to Boston and was operated on several times the past autumn. He was exposed to a harsh easterly wind while taking salt on board his vessel, which caused severe recurrence of inflammation of his right eye. He was not under the care of a physician until the fifth day after the attack. He was bled and then a large portion of cooling cathodics applied, the most mild and cooling wash to the eye. He was kept quiet and secluded from the light. With the inflammation being too far advanced to yield kindly, he took nauseating doses of tartrate. Pain and inflammation gradually abated but the eye tumorfied, which was attended sometimes with a discharge. Two months earlier, he had seton placed in the back of the neck which kept up a copious discharge. On March 23, the day of admission, the right eye was irrecoverably lost, the consequence of ulceration of the cornea. Found was a hernia of the capsule of the aqueous with some of the inner lamina of the cornea. Caustic was applied. He was told to avoid animal food. Fungus growth was diminished by the caustic but a violent inflammation of the conjunctiva occurred. He was given salves. On April 2, he received caustic medication and a wash of the eye. On June 6, the cornea was divided and the membrane detached.

NOW

This patient most likely suffered a corneal surface injury or abrasion ("*exposed to a harsh easterly wind while talking salt on board his vessel*") that was subsequently complicated by an infection of his cornea or keratitis that ultimately resulted in corneal perforation. Infectious keratitis is usually bacterial in nature with *Staphylococci* and *Pseudomonas* species being the most common inciting pathogens. The differential diagnosis of infectious corneal infiltrates includes fungi, parasites (including protozoa such as *Acanthamoeba*), nematodes in developing countries (e.g., *Onchocerca*), and viruses (e.g., *herpes simplex, varicella-zoster, Epstein-Barr*).

With the exception of gonococcal keratitis, bacterial keratitis rarely occurs in an eye in the absence of predisposing factors. Unlike the 1800s when corneal trauma was more common, today, contact lens wear and corneal surgery with local immunosuppression are leading causes of bacterial keratitis and corneal ulceration.

Central corneal scarring and/or topographic irregularity can cause substantial visual loss, even if the bacterial pathogen is successfully eradicated. Untreated bacterial keratitis can result in corneal perforation and endophthalmitis with ultimate loss of the eye, as occurred in this patient in 1822 in the pre-antibiotic era. Because the rate of corneal destruction depends on the virulence of the infecting pathogen (and it can occur in less than 24 hours by highly virulent strains), a prompt diagnosis, initiation of therapy, and follow-up are of utmost importance.

Presenting symptoms of bacterial keratitis include pain, redness, blurred vision, discharge, and photophobia. This patient experienced all of them. Today, this patient would have undergone an ophthalmologic examination that includes measurement of visual acuity, external examination, and, slit-lamp biomicroscopy (developed in the early 1900s). Diagnosis of bacterial keratitis today is based on clinical signs seen with slit-lamp biomicroscopy, such as corneal stromal infiltrates, corneal edema, and ulceration, as well as anterior chamber inflammation that can cause a hypopyon, i.e., layering of white blood cells inferiorly in the anterior chamber.

Corneal smears and cultures can aid in the identification and antibiotic sensitivity of the inciting pathogen and would have also been obtained today during the initial work-up of this patient. Though most cases of community-acquired bacterial keratitis are managed today with empiric antibiotic therapy, corneal smears and cultures are clearly indicated for any infiltrate that is: larger than 1 mm, located centrally in the cornea and thus sight-threatening, chronic or unresponsive to broad spectrum antibiotics, extends into the deep stroma, or is suspicious for a fungal (e.g., ulceration after trauma with vegetable matter), amoebic (e.g., history of wearing contact lenses in a hot tub), or mycobacterial infection. Even in the presence of a hypopyon, aqueous or vitreous taps to rule out endophthalmitis are rarely indicated (e.g., after intraocular surgery, perforating trauma, or systemic sepsis) since the hypopyon due to bacterial keratitis is usually sterile in nature.

The preferred method of treatment for bacterial keratitis is topical broad-spectrum antibiotic eye drops. A loading dose (e.g., every five to fifteen minutes for the first one to three hours), followed by applications every 30 minutes to an hour afterwards, is usually recommended in cases of severe or central keratitis such as seen in this patient, and the preferred agents are either a fourth-generation fluoroquinolone (i.e., moxifloxacin or gatifloxacin) or fortified tobramycin (9-14 mg/mL) and vancomycin (15-50 mg/mL). Cycloplegic agents would also be added in this patient's regimen today in an attempt to decrease pain and to prevent synechiae formation from the anterior chamber inflammation. Ocular ointments can be added to the treatment regimen of bacterial keratitis as adjunctive therapy, though with each additional eye medication compliance has been shown to decrease. Subconjunctival and systemic antibiotics can be helpful in cases of imminent scleral spread or perforation and would have been used in this patient when ulceration was so severe that perforation was inevitable (*"Found was a hernia of the capsule of the aqueous with some of the inner lamina of the cornea"*). Today, application of tissue adhesive, penetrating keratoplasty, and lamellar keratoplasty are used in bacterial keratitis cases complicated by impeding or frank perforation.

Follow-up care for bacterial keratitis depends on the extent and severity of corneal involvement. In severe cases such as this, daily follow-up examinations would be necessary until clinical improvement is seen. Decreased pain, discharge, thickness of stromal edema, anterior chamber inflammation as well as sharper demarcation of the infiltrate, and re-epithelialization are some of the clinical signs and symptoms that indicate a positive response to antibiotic therapy. Topical corticosteroids can be added after at least two to three days of improvement on topical antibiotics and ideally after the infecting organism has been isolated in culture, in order to decrease the inflammatory response and subsequent scar formation of central corneal infiltrates that compromise the visual axis. Tapering of topical antibiotics depends on the severity of infection and rate of improvement. Dosing of topical antibiotics should never be below three or four times a day or the level will become subtherapeutic and pathogen resistance will develop.

If there is no favorable clinical response within 48 hours, the antibiotic regimen is modified. No improvement or stabilization despite change in therapy is usually an indication for reculture – sooner rather than later if the initial culture is negative and an atypical microbial infection is suspected.

Although the majority of patients with bacterial keratitis are treated on an outpatient basis, hospitalization is indicated when the keratitis is vision-threatening, associated with severe pain, or if compliance with the frequent instillation of eye drops is questionable due to the age or physical ability of the patient or his caretakers.

Without antibiotics, the therapeutic choices for physicians in 1822 were limited. Identification of the infecting organism was not possible and treatments were reduced to supportive care and the creation of iatrogenic fistula tracts. Today, in the event of a treatment failure and loss of the eye to infection, microsurgical techniques allow us to remove the eye and implant a prosthesis that makes the patient comfortable and creates an acceptable cosmetic appearance. Two hundred years ago, an eye patch would be the entire extent of oculoplastic reconstruction.

Dr. James Chodosh
Dr. Matthew F. Gardiner
Dr. Joan W. Miller
Dr. Sotiria Palioura
Dr. Athanasios Papakostas

Patient #24: Cancer of the breast

THEN

A married woman, age 49, from Newburyport. She came to the hospital on April 19, 1822, about ten months after she first perceived a small tumor on the right breast, about three inches above the nipple, which gradually increased in size but without pain until the past November. It became discolored. She had acute pain, extending to the axilla. She consulted a physician who thought it was an infection but it became more painful and malignant in its aspect. She was advised to come to the hospital, where upon examination, the tumor appeared to be the size of an egg, reddish-purple, and hard. It was pronounced to be a cancer and extirpation advised. On April 24, all things being ready, an operation commenced by making an incision extending from the axilla to the sternum and a similar one from the same place to the lower sternum, almost in line above the nipple. The tumor was dissected with about one third of the breast, beginning at the sternum and advancing toward the axilla. Several glands had enlarged and indurated. These were taken out (lymph nodes). Three small arteries were secured by ligatures. Little blood was lost. The wound was closed by sutures and adhesive straps. On April 24, she was given tincture of opium and on April 27, the wound was dressed. On May 10, the wound had healed after being dressed every day. On May 15, she was discharged "cured."

NOW

Patients #24 and #37 are examples of locally advanced breast cancer. In the 1800s, nearly all breast cancers presented with large primary tumors, often with visible masses and skin erosion, and frequently with positive axillary nodes. As these records indicate, the most common surgical management was wide local excision of the breast mass, removing a significant portion of the breast and including axillary lymph nodes in continuity. At that time, many clinicians thought that there were strands of tumor connecting the metastatic deposits in the nodes and elsewhere to the primary tumor. Surgical excision therefore included a bridge of tissue between primary tumor and axillary nodes, in hopes of curative resection.

This patient appears to have had a T2 primary tumor with three grossly involved axillary nodes. Given her surgical procedure, she would be expected to have a 75 to 100 percent risk of recurrent tumor in her breast within the next five years, and only a 30 to 40 percent five-year survival rate, due to her significant risk of metastatic disease.

Today, with widespread use of screening mammography, the average size of an invasive breast cancer at diagnosis in Massachusetts is 1.5 cm. Most of these tumors will have negative axillary nodes and survival in the 90 percent or greater range. In addition, 20 to 25 percent of all breast cancers are diagnosed as carcinoma in situ, and have survival in the 99 percent range.

Patients with these early-stage breast cancers are usually managed with breast conserving surgery with lumpectomy alone for carcinoma in situ, and lumpectomy with excision of only one or two sentinel nodes for invasive carcinoma. These surgeries are performed on an outpatient basis with women returning to normal activities in two or three days, and to sports within a week. Most women will receive radiation after mastectomy, although many older women, and some younger women with carcinoma in situ, do not require radiation. Systemic therapy, including endocrine therapy for estrogen receptor positive tumors and chemotherapy for estrogen receptor negative and other more aggressive tumors is used frequently, and can reduce risk of metastasis by up to 50 percent.

For the small number of women who present today with locally advanced breast cancers like this patient, and with no evidence of metastatic disease, it is routine to give systemic therapy, often combination chemotherapy, as the first treatment modality. Following chemotherapy, surgery is performed. Some patients with locally advanced breast cancer become eligible for lumpectomy and axillary dissection if chemotherapy has shrunk their primary tumor sufficiently. For patients whose tumors remain too large for lumpectomy, a modified radical mastectomy is performed, which removes all breast tissue and level I and II axillary nodes. The majority of patients who require mastectomy will ultimately have breast reconstruction. All patients with locally advanced breast cancer receive chest wall and lymph node irradiation after lumpectomy or mastectomy. Patients with endocrine sensitive tumors will receive antiestrogen therapy for five or more years. Local recurrence rates are often five percent or less following these therapies and long-term survival for stage II and III breast cancer is 50 to 80 percent.

Dr. Barbara Lynn Smith

Patient #25: Compound leg fracture

A 56-year-old Scotsman was admitted on April 22, 1822 with a comminuted fracture of his left leg. He had caught his foot and fell over the edge of the wharf. That night he had delirium. He was bled for sixteen ounces and took opium. He was treated with sulfa soda, laudanum, and five days of delirium. Eight days after admission, his delirium was gone. However, on May 8, he died in a stuperous condition. On the postmortem examination, above the fracture he had purulent collection of an undrained abscess.

NOW (See Patient #35)

Patient #26: Death from head injury

THEN

A fourteen-year-old male, an English cabin boy, was brought to the hospital on May 6, 1822. He was unconscious as a consequence of a fall into the hole of the ship, a distance of about fifteen feet. The accident occurred the morning before he was admitted. Examination after death showed the skull to be extensively fractured from traumatic impact, and great violence done to the substance of the brain.

NOW

This patient was dead upon arrival at the hospital.

Patient #27: Venereal ulcers

THEN

A 35-year-old male from Truro was admitted on May 22, 1822. Six years earlier he had had ulcers break out on his arms, one year after having similar ulcers on his legs. The sores had eventually healed. The year before, he was unable to go to sea because of rheumatic changes that had preceded the ulcers and had lasted three to four weeks.

Twelve years earlier he had had gonorrhea and ulcers on his glans penis, which also healed. On June 5, he had received multiple pharmaceutical agents which caused a sore mouth and necessitated a cessation of the medication. The skin ulcers healed well and he was told to wash his sores in salt water twice daily. He was instructed to apply caustics once a month. He was discharged improved.

NOW (See Patient #15)

Patient #28: Elbow swelling

THEN

A 17-year-old woman entered the hospital on May 23, 1822. For the last eleven months she had been troubled with swelling and inflammation above the elbow joint. The last two to three months had been especially painful and she had been unable to use her joint. She received compresses and six leeches applied to the swelling. The pain and inflammation greatly subsided. She received opiates, after which she had perfect use of her arm. Blisters and leeches had been used several times since being in "the house." She was discharged "cured."

NOW

Minor orthopedic condition

Patient #29: Case missing from archives

Patient #30: Leg swelling

THEN

An adult male was admitted to the hospital on May 29, 1822. Two weeks before his arrival by ship he fell on board and was thrown to the deck. As a consequence he developed swollen legs. He was given medication and discharged.

NOW

A patient was admitted with swollen legs after sustaining a fall two weeks earlier while on board a ship. Today, plain films of the pelvis and bilateral extremities would be ordered to rule out fractures. If none were found, lower extremity ultrasound would be used to evaluate the venous system for possible deep venous thrombosis. If clots were not found, and it seemed that the leg swelling was unrelated to the fall, a full work-up would be offered to rule out other causes for the swelling, such as cardiac or renal disease.

Dr. George Velmahos

Patient #31: Genital ulcers

THEN

A 20-year-old male entered "the house" on June 4, 1822, with ulcers on his left leg since February of 1821. The ulcers had been preceded by violent pains in his limb shooting from the foot to the hip. In 1818 or 1819 he had a small ulcer on his glans penis. He took mercury and the ulcer disappeared in three to four weeks. The ulcers on his leg were treated by poultice and opiates. Sulfur powder was placed on the wound. He was discharged on August 3, "cured."

NOW (See Patient #15)

Patient #32: Comminuted fracture of humerus

THEN

A 37-year-old male was admitted June 8, 1822 with comminuted fracture of the left humerus. Splints were applied and he was treated with sulfa soda. He had sixteen ounces of blood removed from his arm. By June 20 the swelling had abated, by June 28 union had occurred, and on July 10 he was discharged.

NOW (See Patient #7)

Patient #33: Ankle sprain

THEN

A 32-year-old female was admitted on June 14, 1822. She had had a violent ankle sprain three years earlier. She was unable to walk for two weeks and then went back to ordinary work. Prior to admission, the foot became swollen and painful. She had twelve leeches applied to the foot.

The rest of her course of treatment included:

6/17 - swelling subsided; bowel stimulation
6/24 - foot slightly swollen – no pain
7/2 - sulfa soda
9/4 - ankle better; create blister
9/24 - dress blister with unguent
10/1 - a poultice of Indian meal and milk
10/3-10/8 - continuous ankle improvement

NOW

Minor orthopedic condition

Patient #34: Injury to skin of foot

THEN

A 55-year-old man was admitted to the hospital on June 25, 1822. He had a bad sore on his right foot, the consequence of striking his foot against a sharp stick of wood seven weeks before. Sulfa soda was used as a dressing, a warm bath and a poultice of white bread and milk daily, as well as other unguents. The sore filled with granulations and he was discharged "cured."

NOW

Today, plain films would be ordered to rule out any foreign material remaining in the foot. Finding none, the patient would be given oral antibiotics and taken to the operating room for debridement of the wound. Because the injury was caused by a wooden object, which may not show on plain films, we would explore the wound intra-operatively for foreign objects. If the wound were very large, we would place a vacuum-assisted closure dressing (VAC®). The patient would be discharged the next day and continue VAC dressing changes with a visiting nurse. The wound would ideally be successfully healed in ten days.

Dr. George Velmahos

Patient #35: Musculoskeletal infection

THEN

A 10-year-old boy came into "the house" with a crutch, able to touch only the toes of his right foot to the floor and unable to bear weight because of ankle weakness. This was caused by a jump and turning of his ankle which had occurred three years earlier. For two years he was able to walk without a crutch, although lame. Then, two abscesses formed near the ankle joint, and they discharged freely for three to four weeks, and then healed. He then walked well until the past spring, when the joint became weak and painful and he was unable to walk without a crutch.

Treatment was started by applying leeches to the joint for three days and then adding vesication. On June 26 leeches and blistering were applied alternatively. On September 27 he received an ankle wash with lead acetate solution with alcohol. The blister on the leg healed but the abscesses appeared to contain pus. Vesication of the ankle discharged a small amount of bloody material. Topical treatments were used until November 1 when the swelling of the ankle appeared to diminish.

On November 19 the wound was probed and the abscess extended below the opening for about an inch in the direction of the internal ankle. There was no communication between the abscesses, although it was noted that the abscess on the external ankle had diminished when the other drained more copiously. Treatments were continued with repeated blistering above and below the joint. By February 12, unguents and ankle washes with sulfa solution were used and the joint appeared perfectly sound and movable without pain. By March 12, the ankle was nearly well, although the right foot and limb were smaller than the other. On April 8 the patient was discharged and considered "cured."

NOW

Infections of the musculoskeletal system can lead to functional deficits, and may be life- or limb-threatening when left untreated. Causes of infections are numerous, and are often associated with local bony or soft-tissue trauma, environmental exposure, or hematogenous spread of a systemic process.

Current treatment options for musculoskeletal infections are multimodal, and often involve a multidisciplinary team including, but not limited to, orthopedic surgeons, infectious disease specialists, radiologists, and wound care nurses. Diagnoses of musculoskeletal infections are made with the use of various types of radiologic imaging, laboratory studies, microbiological data, and physical examination findings. Antibiotics, surgical debridement, and wound care are the keystones of therapeutic intervention.

A review of the earliest documented cases of musculoskeletal infections at the Massachusetts General Hospital provides insight into the diagnostic and therapeutic principles and challenges of that era. Comparing the myriad advancements in technology and knowledge exposes some drastic differences between then and now. However, many of the key treatment principles in musculoskeletal infections are shared between the two eras.

How Far We Have Come

Patient #25 paid the ultimate price of a musculoskeletal infection. This case demonstrates the vast differences in how patients with infections were treated during this era compared to today. It is unclear whether his infection preceded his fracture or vice versa. He was treated with analgesics (laudanum, aka tincture of opium) and bloodletting, a common practice during that time period but virtually abandoned during modern treatment of infections.[1] Ultimately, the Scotsman suffered from a lack of diagnosis because the abscess in his leg was not discovered until postmortem. Today, he would have received radiographs of his fracture, and likely an MRI or CT scan to assess the soft tissues, and search for a drainable collection. Laboratory data would include a white blood cell count with differential, and inflammatory markers such as an erythrocyte sedimentation rate (ESR) and C-Reactive Protein (CRP).[2] His initial work-up would begin in the emergency room, and he would likely be admitted to the Orthopedic Trauma Service. He would undergo surgical debridement, fixation of the fracture, and intravenous antibiotics would be administered. Culture data and consultation from an infectious disease specialist would orchestrate antibiotic choice and duration. He would likely survive his illness with little to no morbidity.

The 10-year-old boy described as Patient #35 likely had a chronic osteomyelitis and/or septic arthritis of the ankle, with resultant draining sinus tracts. He was successfully treated with local wound care. Like Patient #25, if this boy were to present today, he would undergo a thorough diagnostic evaluation and probable surgical debridement. Although leech therapy is still utilized to treat venous congestion in microsurgery,[3] it would not be implemented in a case like this. Notably, his treatment did not include antibiotics, since they had not yet been discovered. Today, antibiotics are the mainstay, and often the only form of treatment necessary for osteomyelitis.[2]

Infections of the spine are among the most debilitating musculoskeletal infections, given their depth in the body and proximity to the neural elements. The lumbar spine is most commonly involved.[]

Abscesses and osteomyelitis can form in the osseous structures or epidural and intradural spaces, occasionally leading to neurologic insult.[5] Patient #40, a teenage girl, had some variation of a lumbar spine infection, though it is not entirely clear from her chart. Her treatment consisted of supportive care with no clear treatment directed at the infection itself. Ipecac, an expectorant and emetic, was used presumably to prevent bowel obstruction or to purge the illness. Emetics are seldom used today, and currently have no role in the treatment of infections of the spine.[6] Similar to Patients #25 and #35, Patient #40 would undergo a thorough laboratory and radiographic evaluation to ascertain the location and extent of her "abscess." Advanced cross-sectional imaging (MRI) would determine if the condition could be treated with antibiotics alone, or if surgical drainage would be necessary. CT or a fluoroscopically guided biopsy by an interventional radiologist might be utilized to reveal a causative pathogen to provide information to the infectious disease specialist, who would determine antibiotic selection and duration.

The most apparent treatment lacking in these three cases is the use of antibiotics, which was not implemented until the World War II era.[7] Today, antibiotics are arguably the single most important tool used to treat infections, and are the common treatment of nearly all infections today. There are currently more than 100 different antibiotics available to treating clinicians.[8]

Standing the Test of Time

Review of the earliest cases at MGH demonstrates an early understanding that surgical debridement was essential to the eradication of musculoskeletal infections. Despite the armamentarium of powerful antibiotic medications that exist today, the importance of surgical debridement remains.

Patient #53, likely a miner or construction worker, sustained a proximal humerus fracture that went on to form an infected non-union. Though the chronology of his treatment is not entirely clear, it appears that he was treated with debridement of nonviable bone and soft tissue, followed by subsequent packing (lint) and serial dressing changes. Eventually, his fracture united and his wounds healed by secondary intention. Although there have been recent advancements in therapies to facilitate secondary wound healing, such as negative pressure wound therapy,[9] the treatment strategy utilized during this case closely resembles that of a case during the modern era.

The treatment of Patient #75 further illustrates similarities between the treatment of musculoskeletal infections during the 1820s and today. This patient had osteomyelitis with ulceration, sequestrum (dead bone), and involucrum (new bone) formation. He was treated with serial debridement and irrigations, as well as packing of his wounds. The necrotic portions of bone were debrided until clean, healthy bone was encountered, and his wounds were irrigated with magnesium sulfate. Antibiotics, advancements in surgical equipment and irrigation solutions notwithstanding,[10,11] this patient underwent remarkably similar management when compared to today.

One major difference highlighted by these two cases, however, is that they were treated in an era without anesthesia. Patient #75 was "held down by assistant," a far cry from the modern methods utilized by anesthesiologists today. Additionally, in today's modern methods utilized by hospitals and health care systems, there is an emphasis on minimizing length of stay.[12] Patients #53 and #75 were admitted for nearly four and five months, respectively, for their infections. Patients with these same predicaments today would likely be hospitalized for five days at most.

Despite the advancements made in the treatment of musculoskeletal infections over nearly 200 years, researchers and providers at the MGH and hospitals throughout the world continue to make improvements, while facing the next generation of challenges ahead.

References

1. Jhang JSJ, Schwartz JJ. Phlebotomy or bloodletting: from tradition to evidence-based medicine. Transfusion. 2012 Mar. 1;52(3):460–2.

2. Lew DP, Waldvogel FA. Osteomyelitis. New England Journal of Medicine. Mass Medical Soc; 1997;336(14):999–1007.

3. Green PA, Shafritz AB. Medicinal Leech Use in Microsurgery. YJHSU. Elsevier Inc; 2010 Jun. 1;35(6):1019–21.

4. An HS, Seldomridge JA. Spinal Infections. Clinical Orthopaedics and Related Research. 2006 Mar.;443(&NA;):27–33.

5. Bluman EM, Palumbo MA, Lucas PR. Spinal Epidural Abscess in Adults. Journal of the American Academy of Orthopaedic Surgeons. 2004 May 1;12(3):155–63.

6. Lee M. Ipecacuanha: the South American vomiting root. The journal of the Royal College of Physicians of Edinburgh. 2008;38(4):355.

7. Waksman SAS, Woodruff HBH. Selective Antibiotic Action of Various Substances of Microbial Origin. J Bacteriol. 1942 Sep. 1;44(3):373–84.

8. Beers MH, Porter RS. The Merck Manual of Diagnosis and Therapy 18th Edition. 2006. p. 2991.

9. Braakenburg A, Obdeijn MC, Feitz R, van Rooij IALM, van Griethuysen AJ, Klinkenbijl JHG. The Clinical Efficacy and Cost Effectiveness of the Vacuum-Assisted Closure Technique in the Management of Acute and Chronic Wounds: A Randomized Controlled Trial. Plastic and Reconstructive Surgery. 2006 Aug.;118(2):390–7.

10. Anglen JOJ. Wound irrigation in musculoskeletal injury. J Am Acad Orthop Surg. 2001 Jan. 1;9(4):219–26.

11. Brown NMN, Cipriano CAC, Moric MM, Sporer SMS, Valle Della CJC. Dilute betadine lavage before closure for the prevention of acute postoperative deep periprosthetic joint infection. Journal Of Arthroplasty. 2012 Jan. 1;27(1):27–30.

12. Kominski GF, Witsberger C. Trends in length of stay for Medicare patients: 1979-87. Health Care Financ Rev. 1993 Jan. 1;15(2):121–35.

Dr. Michael D. Baratz
Dr. Harry E. Rubash

Patient #36: Urethral stricture

THEN

A male was admitted on July 10, 1822, with disease of the urethra. Nineteen years ago he fell outside and punctured his urethra and formed an abscess in the perineum, which drained at that time. He had a suprapubic puncture. When the abscess healed the suprapubic tube was removed, but he then developed a urethral stricture and had to use a catheter. He had not used it for one year, however. On July 15 he was voiding every fifteen minutes in small amounts, and was dilated with a bougie, somewhat relieved. On July 21 he was discharged.

NOW (See Patient # 4)

Patient #37: Cancer of the breast

THEN

A 51-year-old woman was admitted on July 16, 1822 with a cancerous breast. Two months earlier she noticed indurations and drainage in the skin in the upper part of her left breast. Extirpation was advised. She was treated pre-operatively with sulfa soda.

An operation was performed on July 17, with one third of the breast removed above nipple and closed with adhesive straps. On July 24 there was no infection and she was healing well. She was discharged "cured" on July 30.

NOW

Patients #24 and #37 are examples of locally advanced breast cancer. In 1800s, nearly all breast cancers presented with large primary tumors, often with visible masses and skin erosion, and frequently with positive axillary nodes. As these records indicate, the most common surgical management was wide local excision of the breast mass, removing a significant portion of the breast and including axillary lymph nodes in continuity. At that time, many clinicians thought that there were strands of tumor connecting the metastatic deposits in the nodes and elsewhere to the primary tumor. Surgical excision therefore included a bridge of tissue between primary tumor and axillary nodes, in hopes of curative resection.

Patient #37 had an even more advanced T4 tumor, with direct skin involvement and ulceration. No mention is made of lymph node involvement, but this would be very likely. She also would have a 75 to 100 percent of early local recurrence, and at best a 25 to 30 percent five-year chance of survival.

Today, with widespread use of screening mammography, the average size of an invasive breast cancer at diagnosis in Massachusetts is 1.5 cm. Most of these tumors will have negative axillary nodes and survival in the 90 percent or greater range. In addition, 20 to 25 percent of all breast cancers are diagnosed as carcinoma in situ, and have survival in the 99 percent range.

Patients with these early-stage breast cancers are usually managed with breast conserving surgery with lumpectomy alone for carcinoma in situ, and lumpectomy with excision of only one or two sentinel nodes for invasive carcinoma. These surgeries are performed on an outpatient basis with women returning to normal activities in two or three days, and to sports within a week. Most women will receive radiation after mastectomy, although many older women, and some younger women with carcinoma in situ, do not require radiation. Systemic therapy, including endocrine therapy for estrogen receptor positive tumors and chemotherapy for estrogen receptor negative and other more aggressive tumors is used frequently, and can reduce risk of metastasis by up to 50 percent.

For the small number of women who present today with locally advanced breast cancers like those of Patients #24 and #37, and have no evidence of metastatic disease, it is routine to give systemic therapy, often combination chemotherapy, as the first treatment modality. Following chemotherapy, surgery is performed. Some patients with locally advanced breast cancer become eligible for lumpectomy and axillary dissection if chemotherapy has shrunk their primary tumor sufficiently. For patients whose tumors remain too large for lumpectomy, a modified radical mastectomy is performed, which removes all breast tissue and level I and II axillary nodes. The majority of patients who require mastectomy will ultimately have breast reconstruction. All patients with locally advanced breast cancer receive chest wall and lymph node irradiation after lumpectomy or mastectomy. Patients with endocrine sensitive tumors will receive antiestrogen therapy for five or more years. Local recurrence rates are often five percent or less following these therapies and long-term survival for stage II and III breast cancer is 50 to 80 percent.

Dr. Barbara Lynn Smith

Patient #38: Case missing from archives

Patient #39: Bilateral patella fracture

THEN

A 21-year-old male was admitted on August 2, 1822. Riding in a wagon the day before, he received kicks on his knees from an unruly horse. A surgeon was called in and confirmed fractures of both patella. A portion of the right bone was removed. He was brought to the hospital at 6 p.m. Both knees were considerably swollen, but not painful. He bled freely, got sulfa soda immediately, and applied cloth to the knees with wet solutions. His knees became painful, particularly at night. He got hydrangea and opii three times a day, twelve leeches applied to the right knee.

On August 4, the right knee was less swollen. Wounds were dressed with simple ointment. On August 6, there was no pain or swelling in the left knee but the right knee was swollen and inflamed. Leeches and poultice of oatmeal were given, as well as opium every night. The right knee was improved on August 10. There was separation of the joint and free discharge from the wound where the piece was removed. The other side of the knee was opened with a lancet. On August 13, the knee was much improved. Poultice was used, took opium pill, and four calomel pills daily. His mouth was sore on the August 16 so the pills were omitted. The right knee was more painful and they applied a fomentation of wormwood and tansy. There was increased swelling of the knee on August 28. On August 31, he got tincture of cinchona, three times a day, to his knee. On September 2, the ankle was more painful, knee was washed with alcohol. On September 7, the discharge from the right knee was greatly diminished, but it continued to swell. It was dressed twice a day.

On October 1, he was continuing to heal. He did better with silver nitrate and the knee was irrigated daily, applying the liquid solution. On October 18, he complained of pain in his head and soreness in the limbs. The previous night, he took no medications. He had nausea and pain in his head and took ipecac and aqua patina, and other fomentations were applied to the right knee. The next day, he complained of general soreness of muscles, his pulse was 116 and his tongue was dry. He received compresses repeated every four hours. On October 17, the wound appeared healing, swelling was nearly gone. Pulse was down to 90. He slept without pain on October 18 and he was discharged "cured" on October 19.

NOW

The case above describes a young male patient who sustained direct trauma to his bilateral knees from a horse. It appears the horse kicked him directly over the anterior knees causing fractures, one of which appears to be open (meaning the fracture is exposed through the skin as illustrated in the proceeding examples of open fractures). The patient was initially evaluated by a surgeon, who removed a portion of exposed bone, a procedure that is akin to modern day irrigation and debridement. This is carried out for open fractures or injury where devitalized, necrotic, or contaminated tissue is removed.

Given the extent of the injury, the patient was then transferred to the hospital for evaluation. His initial treatment consisted of being "bled freely" and application of sulfa soda with warm dressings. Allowing the wound to drain, or "bleed freely," helps to further clear contamination and create a healthy wound healing environment. Sulfa soda and warm compresses were used as an early antibacterial to aid in preventing infection. Sulfur in sulfa soda was believed to have antimicrobial properties and many early antibiotics were developed due to this including the sulfonamides, penicillins, and cephalosporins (all of which contain sulfur compounds).[1]

The patient's knees continued to be painful and he received opium for pain control in addition to receiving hydrangea. Hydrangea, which is a common flowering plant, was believed to have medicinal uses and was often used as a diuretic, cathartic, or even a pain reliever. It is also purported to have anti-malarial activity.[2] In addition to this therapy, leeches were applied to the open wound on the right patella. The medical leech, or what is scientifically referred to as *Hirudo medicinalis*, has been used for clinical bloodletting for thousands of years, the earliest documentation being in India. In ancient Greece, bloodletting was practiced according to the humoral theory. The Greeks believed that the body consisted of four "humors" (blood, phlegm, black bile, and yellow bile). When these humors were in balance, good health was guaranteed. Conversely, an imbalance in the proportions of these humors was believed to be the cause of illnesses. Bloodletting using leeches was one method physicians used to balance the humors in an attempt to control disease.

Modern medicine uses leeches in several applications. Leeches secrete an active anti-coagulant, known as hirudin, in their saliva. This anticoagulant activity is useful for promoting blood flow to damaged or injured areas. Leeches are utilized in vascular surgery applications to avoid venous congestion in tenuous tissues. They also serve to engulf bacteria and other contaminants in the wound bed. This, in addition to the anticoagulant and blood flow effects, aids in tissue healing.[3] We see this illustrated in the case above where the leeches aided in reducing swelling of the knee. Oatmeal poultice was also described as a treatment, which essentially serves as a heat pack for the area. Hot oats are spread onto a piece of linen and strapped to the body using a bandage. The warmth increases blood flow to the affected area, soothing skin irritation and promoting healing, theoretically having a compounding effect along with the leeches.

Despite these therapies, the case notes continued drainage from the area with the joint described as "separated." Although this cannot be proven, it appears that there was a patella fracture with complete disruption of the bone and skin with displacement of the fracture pieces, which allows direct visualization into the knee joint itself. It also appears that an abscess had formed and needed to be surgically opened with a lancet to allow drainage of infection. In a modern day situation, this patient would be admitted to the hospital, have a local irrigation completed at the bedside to evacuate contamination, and be started on antibiotics. He would be taken to the operating room and have the fracture fixed and the skin closed. Antibiotics would be continued and the wound would be closely monitored for signs of infection. Again, an aggressive approach is taken with open fractures.

After this patient had a small local debridement (lancet opening the other side of the knee), conservative therapy was continued with calomel pills, poultice, and fomentation of wormwood and tansy. Calomel pills were classically used to treat eruptive diseases of the skin, rheumatism, hepatitis, and assorted liver ailments. Calomel pills are a combination of sugar, molasses, syrup, and a sulfur-based form of the metal antimony. The problem is that the body ultimately changes the sulfured form of antimony to a sulfured form of mercury. Mercury is toxic to human tissues and inhibits several enzymes throughout the human body.[4]

Common symptoms of mercury poisoning include peripheral neuropathy, or burning pain, skin discoloration, fast heart rate, sweating, high blood pressure, swelling, and desquamation (shedding of skin). A very common feature of mercury toxicity is seen around the face and mouth and patients may exhibit mouth pain; red cheeks, nose, and lips; or loss of hair, teeth, and nails.[5] Given the statement, "his mouth was sore on the 16th and so the pills were omitted," the patient above seemed to show signs of early mercury poisoning.

Also described are wormwood and tansy (herbs and plant substances), which have traditionally been used in remedies to reduce fevers, soothe the stomach and digestive tract, and to treat gout. Externally, in addition to poultices, it has also been used to treat some skin infections.

The physicians tried diligently to prevent the progression of this patient's problem but unfortunately he continued to experience pain and swelling. New therapies with tincture of cinchona, alcohol, and silver nitrate were attempted. Cinchona, which is a flowering plant species, was used as an antiseptic and astringent. A poultice of the bark has been successfully used on gangrenous ulcers to allow suppuration. When taken by mouth, the ingredients in cinchona were believed to be active against malaria.[6] Silver nitrate (an inorganic silver compound) was traditionally used as an antimicrobial agent, used topically to treat superficial infection. With the advent of modern day antibiotics and topical therapy, it has fallen out of favor.

Although the patient's wound and swelling continued to improve, it appears that he became systemically ill. He had nausea, vomiting, dehydration (mouth was dry), headache, general lethargy, muscle soreness, and malaise. This could be secondary to a systemic infection from his knee but given the limited information in the case presentation, it is hard to elucidate further. He received ipecac, which is an alcoholic extract of the roots of the flowering plant *Carapichea ipecacuanha*. Ipecac root itself is a poison, but in its diluted form (ipecac solution) it is used to induce immediate vomiting. Ipecac solution is a mixture of this diluted root solution, simple syrup and sugar.[7] In a dehydrated patient the induction of vomiting is not recommended and can lead to further dehydration. Fortunately, the patient continued to improve and was discharged as "cured."

This case is a good example of the therapies at the disposal of physicians to treat common orthopedic ailments in the early nineteenth century. If this patient were to be seen at the Massachusetts General Hospital today, he would be evaluated in the emergency room, have his wounds cleaned, his fracture fixed, and antibiotics administered over a few days. A typical hospital stay following this type of injury would be three to four days. This patient, however, spent almost 80 days in the hospital. It is amazing to see how technology and innovation have changed medicine and we are excited to see what the future has to bring.

References

1) Aguirre, C. Antibiotics, sulfa and nitrofurans drugs. Cir Cir. 1957. Apr; 25(4): 201-5.

2) Kamei K, Matsuoka H, Furuhata SI, Fujisaki RI, Kawakami T, Mogi S, Yoshihara H, Aoki N, Ishii A, Shibuya T. Anti-malarial activity of leaf-extract of hydrangea macrophylla, a common Japanese plant. Acta Med Okayama. 2000 Oct; 54(5): 227-32.

3) Wiwanitkit V. Leech therapy. Anc Sci Life. 2012 Jan; 31(3): 141.

4) Norn S, Permin H, Kruse E, Kruse PR. Mercury--a major agent in the history of medicine and alchemy. Dan Medicinhist Arbog. 2008; 36:21-40.

5) Tezer H, Kaya A, Kalkan G, Erkocoglu M, Ozturk K, Buyuktasli M. Mercury poisoning: a diagnostic challenge. Pediatric Emergency Care. Nov; 28, 2012. (11): 1236-7.

6) Skogman ME, Kujala J, Busygin I, Leinob R, Vuorela PM, Fallarero A. Evaluation of antibacterial and anti-biofilm activities of cinchona alkaloid derivatives against Staphylococcus aureus.
Nat Prod Commun. 2012 Sep; 7 (9): 1173-6.

7) Lee MR. Ipecacuanha: the South American vomiting root.
Journal of the Royal College of Physicians Edinburgh. 2008 Dec; 38(4): 355-60.

Dr. Jeffrey R. Jaglowski
Dr. Harry E. Rubash

Patient #40: Muscular skeletal infection (psoas abscess)

THEN

A 17-year-old woman was admitted on August 6, 1822. Diagnosis was lumbar abscess. She got ipecac. Her history showed that she had a fall from a wall the previous winter and something jarred her whole system but she experienced no particular inconvenience directly after the incident. After two months, she had a fixed sharp pain in the side. The pain became permanent and more severe and extended around to her back to the other side. The pain was darting and was so great it prevented her from walking. She was blistered before she entered the hospital. The pain in the side was very much diminished and the lameness by a great degree was gone by September 22, previous to discharge. On September 24, she had pain on the right side with some feverish symptoms. Pulse was 116. On October 2, her pulse was 110. She was well until October 4 when she complained of headache and severe nausea which continued through the day. On October 5, her face was slightly flushed, headache was gone, she had vomited undigested food the night before, no discharge from bowels, and her pulse was 112. On October 25, she was given directions to pursue treatment to keep the tissue open, take gentle exercise and open air, and avoid fatigue.

NOW (See Patient #35)

Patient #41: Squamous cell carcinoma of vagina

THEN

A 23-year-old female was admitted on August 12, 1822, complaining of anuria for two to three days. Examination shows gangrene in the vagina with symptoms of vomiting, sore throat. The patient was treated with hippac, opium, and muriatic acid. Mercury factor in her breath from pills taking before coming to the hospital. On August 15 the patient died. The cause of the gangrene was unknown.

NOW

This woman's history was largely obtained through her friend since she was incoherent. The patient was born in Africa and she was sold by her family at the age of ten. She served as a teenage prostitute for the next decade prior to being emancipated and fleeing to America at age of 20. She worked at various housekeeping jobs for a few years and had noted increasing vaginal discharge. She had purchased some over-the-counter vaginal yeast suppositories without improvement. She did not have any health insurance and did not seek medical care. She recently moved to Massachusetts where she qualified for universal coverage via the MassHealth plan, but had not yet applied. Over the past few months her vaginal discharge had become increasingly malodorous. She was taking several ibuprofen pills on a daily basis due to severe pelvic cramps and back pain. She had not felt well for the past two weeks and was brought to the ED by her roommate because she had been increasingly somnolent and this morning was not arousable.

The ED triage nurse took her vital signs: heart rate 130 beats per minute, blood pressure 70/35, temperature 99.0 Fahrenheit, 30 respirations per minute, oxygen saturation 90% on room air. She was admitted to the ED, where an intravenous (IV) line was difficult to place due to her collapsed veins. Vigorous IV hydration was initiated. Supplemental oxygen was provided via face-mask. The patient was combative and required restraints. She was experiencing dry heaves. Routine labs were drawn and the ED attending physician was consulted. Most notably, her creatinine was 12 [normal range, 0.6-1.2] and her potassium level 7.0 [normal range, 3.5-5]. Her other electrolyte abnormalities were less impressive. Her hemoglobin level was 6 [normal 10-12] and white count elevated at 18,000 [normal 5-11,000]. She was placed on telemetry where brief runs of atrial fibrillation were seen. A foley catheter was placed. Minimal urine was noted, but necrotic tissue was observed at the vaginal introitus. The ED team requested a consult from the gynecology service.

The on-call gynecology resident performed a bedside exam that was limited by the patient's pain and inability to cooperate due to her altered mental status. Necrotic tissue was seen to fill the upper vaginal vault and a sample was sent for pathology review. A bedside renal sonogram was requested. The radiologist arrived in the ED and a quick scan showed severe bilateral hydronephrosis.

The patient's vital signs were improved with vigorous hydration. The ED attending placed a larger venous access device and requested a renal consult. She continued to be aggressively hydrated via both her peripheral intravenous line and central venous access line. She was transferred to the dialysis unit and underwent semi-emergent hemodialysis. The patient was seen by the medicine service and admitted as an inpatient once her hemodynamic status had been further stabilized. Interventional radiology was consulted and bilateral percutaneous nephrostomy (PCN) tubes were inserted. Her creatinine improved to 6.5 and copious urine was observed entering the drainage bags. The medicine team continued to manage her electrolyte abnormalities over the next few days as her mental status dramatically improved.

The gynecology resident was paged from pathology that the vaginal tissue was consistent with squamous cell carcinoma. A gynecologic oncology consultation was requested. A non-contrast CT scan was obtained and demonstrated an 8 cm cervical mass with borderline pelvic adenopathy. She was taken to the operating room for a cystoscopy, exam under anesthesia, and proctoscopy. Based on that clinical exam, she was assigned a diagnosis of International Federation of Obstetrics and Gynecology [FIGO] stage IIIB cervical cancer. The medicine team confirmed that she was HIV-positive. She was transferred from the medicine service to the Vincent Gynecologic Oncology Service on Bigelow-7.

She was seen by case management who arranged for her to file for MassHealth insurance through the state of Massachusetts – one of the few states in the country to provide such universal coverage. The nursing staff provided teaching on how to empty the PCN drainage bags. Physical therapy was consulted to assess her ability to function outside the hospital. The infectious disease team was consulted and drew additional blood work to assess the severity of her HIV disease. She resumed her ability to tolerate oral medications and was eating normally. She was started on highly active antiretroviral therapy (HAART) and discharged to her Cambridge apartment on November 27 in the company of her friends. She was given an appointment with her gynecologic oncologist within a few days.

On November 28, the Vincent OB/GYN inpatient nurse practitioner called the patient at home to assess her transition and determine whether she was managing her medications and symptoms. She reiterated the need for follow-up. On November 30, the patient's case was presented at the Multidisciplinary Gynecologic Oncology Tumor Board on Founders-4 in the Joseph V. Meigs Room. Her pathology was reviewed before an assembled 30-person team of gynecologic oncologists, medical gynecologic oncologists, radiologists, case management, gynecologic radiation oncologists, nursing, Harvard Medical Students and Mass General OB/GYN residents. Her care was discussed and an evidence-based consensus was achieved: she would be counseled her about participation in a collaborative group clinical trial with the intent to thereafter recommend 45 Gray of pelvic radiation with weekly cisplatin sensitizing chemotherapy.

She was seen in the Yawkey Center outpatient gynecologic oncology office on December 3. Her lab work had normalized with asymptomatic anemia and a creatinine of 1.6. She was counseled and elected to enroll in the Gynecologic Oncology Group protocol #233 trial. A study-indicated positron emission tomography (PET) scan on December 5 was equivocal for whether her pelvic adenopathy represented metastatic disease. She underwent a CT-guided biopsy on December 7 in the interventional radiology suite and final pathology did not demonstrate squamous cell carcinoma. Her films were jointly reviewed by the radiologist and gynecologic oncologist. She underwent laparoscopic pelvic and para-aortic lymphadenectomy as a day surgery procedure on December 10 and no metastases were detected. A pelvic magnetic resonance imaging (MRI) test was obtained December 13 to better define her pelvic disease and help guide the radiation treatment planning.

She was simulated for radiotherapy on December 14 and began pelvic radiation on Monday December 19 with daily doses of 180 cGy along with weekly cisplatin therapy. Her pain improved during therapy and she no longer required narcotic medication. She had ureteral stents placed by urology and the PCN tubes were removed. She completed 45 Gy and received an additional 900 cGy boost. Her cervical anatomy was vastly improved and she completed an additional course of high-dose rate (HDR) brachytherapy.

Three months after her initial ED admission, the patient was clinically free of disease. She was evaluated by urology and stents were removed. She returned to work and a post-treatment PET scan three months after completion of chemo-radiation did not detect any evidence for persistent disease. She is seen quarterly for a joint surveillance visit with gynecologic oncology and radiation oncology. Here HAART medications have been revised and her CD4 viral counts are undetectable.

Dr. John O. Schorge

Patient #42: Lumbar herniated disc

THEN

A 30-year-old male from Andover was admitted on August 6, 1822. Three months previously, on a journey, he sprained his leg getting out of a carriage. He didn't think it was serious, but after 24 hours he experienced severe pain in the hip which extended to the whole limb. This lasted for about five days, until he arrived at the end of his journey. The pain was so great for the five days that he was unable to use the limb, and then for ten to twelve days he could only move with the greatest difficulty. He had considerable pain in the lumbar region. He soon got better and returned home. The left side of the leg was particularly affected. In February, the leg was blistered and again in April. He returned in May, walked a great deal, often six or seven miles. He only had some stiffening of the limbs upon setting out. For the last three or four weeks before admission, he had darting pains to the hip and to the groin, great instability in the leg, and frequently prickling sensation extending from the foot and toes. In a short time, he had used leeches and liniments without any relief. A blister was made over the peroneal nerve on the leg. On August 13, he was discharged.

NOW

The medical record describes a 30-year-old man with left lower back pain and symptoms of pain and tingling that traveled down his leg and into his toes. This patient likely had a lumbar disc herniation with radiculopathy or nerve pain down the leg. The man received localized treatment to the peroneal nerve, a nerve originating in the lumbar spine, which courses on the outside of the lower leg and provides motor innervation to the anterior and lateral leg muscles and sensation to the top of the foot. The man's symptoms were intermittent, first appearing in August, 1822, and subsequently improving before worsening again in February, April, and August, 1823, one year from the initial onset. While the therapies that were provided for the gentleman would be considered unusual by current standards, the story of a patient with a lumbar disc herniation with radiculopathy is commonly seen at Massachusetts General Hospital today.

Lumbar intervertebral disc herniation is extremely common with more than 200,000 surgeries performed each year in the United States for this diagnosis (Lee, Amorosa, Cho, Weidenbaum, & Kim, 2010). The intervertebral disc is an important part of the lumbar spine, providing both structural support and allowing mobility. The intervertebral disc is an avascular structure lying between two adjacent vertebral bodies. Each intervertebral disc is composed of two distinct regions: an inner gelatinous nucleus pulposus providing compressive strength, and an outer annulus fibrosis providing tensile strength. As the intervertebral disc ages, the inner nucleus desiccates and loses some compressive strength and the annulus can develop fissures and cracks (Lee et al., 2010). With continued axial loading of the lumbar spine, the inner gelatinous nucleus can herniate through the outer annulus and cause compression of a lumbar nerve root. The most common site for this to occur is between the fourth and fifth lumbar vertebrae (L4-5 disc), which commonly compresses the fifth lumbar nerve (L5 nerve root) (Lee et al., 2010). The L5 nerve root is the origin of many of the peroneal nerve fibers; therefore, patients with a L4-5 disc herniation often experience weakness and pain in the distribution of the peroneal nerve, similar to this patient from 1822.

We will never definitively know this patient's diagnosis, but based on the clinical history obtained from the MGH medical records, it appears that he likely had an L4-5 disc herniation with a L5 nerve radiculopathy. A patient with similar signs and symptoms presenting today to MGH would, after a trial of conservative management, likely undergo an MRI, the diagnostic study of choice for lumbar disc herniation. In 1822, the man was treated for an entire year with blistering of his skin, liniments, and leeches. Treatment for lumbar disc herniation today remains a matter of debate and includes both operative and non-operative options. Patients with an initial presentation of a lumbar disc herniation are offered a trial of non-operative therapy including activity modification, non-steroidal anti-inflammatory medications, physical therapy, and possibly epidural or selective nerve root corticosteroid injections (Lee et al., 2010). Should the patient's symptoms fail to improve after several months of non-operative therapy, the patient could be considered for surgical intervention, typically a partial laminotomy and discectomy.

Recently, a large multicenter randomized trial was conducted to investigate the comparative effectiveness of surgical versus non-operative therapy for patients with lumbar intervertebral disc herniation and radiculopathy. The Spine Outcomes Research Trial, or SPORT, randomized 501 patients with lumbar disc herniation at thirteen spine centers across eleven states to either operative or non-operative therapy and followed the patients for four years. The results of the study indicate that patients undergoing surgery for a lumbar disc herniation achieve greater improvement than non-operatively treated patients in all tested outcomes except work status (Weinstein et al., 2008). Interestingly, a quarter of a century earlier, in 1982, Dr. Henrik Weber of Norway presented nearly identical results. His series of 126 patients also demonstrates that operatively treated patients with lumbar disc herniation do better when compared to conservative therapy, although the effect size seen is greater at a one-year follow-up and diminishes at four years (Weber, 1983).

This patient was treated for over a year without resolution and therapies were based on the medical knowledge of the time. His skin was blistered and leeches were applied to the site of his pain, the peroneal nerve. However, it now seems clear from the history presented in the medical records that the source of his symptoms was likely his lower back and not his leg. The same patient presenting to MGH today would likely undergo a trial of non-operative therapy followed by an MRI of his lumbar spine should his symptoms persist. The MRI would likely reveal a lumbar intervertebral disc herniation and a discussion would occur between the patient and his treating physician about the risks and benefits of operative versus non-operative treatments.

References

Lee, J. K., Amorosa, L., Cho, S. K., Weidenbaum, M., & Kim, Y. (2010). Recurrent lumbar disk herniation. *The Journal of the American Academy of Orthopaedic Surgeons*, *18*(6), 327–337.

Weber, H. (1983). Lumbar disc herniation. A controlled, prospective study with ten years of observation. *Spine*, *8*(2), 131–140.

Weinstein, J. N., Lurie, J. D., Tosteson, T. D., Tosteson, A. N. A., Blood, E. A., Abdu, W. A., et al. (2008). Surgical Versus Nonoperative Treatment for Lumbar Disc Herniation. *Spine*, *33*(25), 2789–2800. doi:10.1097/BRS.0b013e31818ed8f4

Dr. Stephen T. Gardner
Dr. Harry E. Rubash

Patient #43: Musculoskeletal inflammation

THEN

A 14-year-old boy from Danvers came into the hospital on August 26, 1822. Three years earlier he had a violent pain in the right hip, darting down into the leg. The limb became shorter, and the boy now walked with crutches. There was contraction of muscles in the thigh, which caused him to be unable to extend the limb by 45 degrees. The previous winter, two or three ulcers opened on the thigh which continued. The judgment was that there was no possibility that the limb could be restored. The patient wished to remain a few days. On September 7, blisters were made but there was no relief. On the August 27, the limb was then noticed to be shortened, wash was applied over the hip and vesication was done. Pain in the hip was much relieved as well as the knee. On October 5, he had soreness around the blister. Blister was three inches square below the hip, blisters kept being applied. On October 15 the blister was healed. On October 19, he had severe pain in his hip since morning; fomentation was applied to the hip for one half hour. There was noted to be a collection of matter under the fascia of the thigh and about the joint. The right limb was shortened considerably; the right hip was much swelled and very hot, with his bowel loose. A blister was applied to the hip. The right limb was two inches shorter than the left. Poultice was placed. On November 4, bowels were active. On November 17, he was finally discharged as incurable. There was no probability of the limb being restored.

NOW (See Patient #35)

Patient #44: Foreign body in eye

THEN

A 25-year-old male was admitted on August 31, 1822. While working as a blacksmith, an iron particle passed through his eye. The sclerotic area and the particle were removed by making an incision through the lateral portion of the cornea where the aqueous humor was discharging. Forceps were used to pick out the fragments. Extract of belladonna was placed around the inflamed eye. On September 2, 1822, the patient was discharged with directions to continue washing the eye.

NOW

This patient had an intraocular metallic foreign body. For these patients, initially a complete exam should be done with special attention paid to the intraocular pressure and fundus to look for other, deeper foreign bodies. The entry site should be carefully examined to check for actively leaking aqueous. Gentle gonioscopy may be needed to look for additional foreign material in the anterior chamber angle of the eye. A CT scan of the eye and orbit is commonly done as well. Treatment includes surgical exploration and removal of the foreign body in the OR. Patients receive IV vancomycin and cefazolin and consequently would undergo removal of the foreign bodies in the OR with suturing of the corneal laceration as well as any other lacerations found. If the patient has signs of endophthalmitis, treatment with intravitreal antibiotics (vancomycin and ceftazidime) is indicated.

It is impressive that the eye surgeons of 1822 were able to remove corneal foreign bodies without the aid of an operating microscope, microsurgical instruments, or proper asnesthesia. The level of discomfort likely experienced by patients undergoing such procedures was certainly extraordinary.

Dr. James Chodosh
Dr. Matthew F. Gardiner
Dr. Joan W. Miller
Dr. Sotiria Palioura
Dr. Athanasios Papakostas

Patient #45: Breast tumor

THEN

A 30-year-old married woman was admitted on September 22, 1822. She had a year of breast soreness. A tumor was found in the right breast the size of a walnut. She suffered lancinating pain. On September 17 her pain and soreness increased. On September 20, almost the entire right breast was removed without much bleeding. Straps were attached. She suffered recurrent wound bleeding.

NOW

Patient #45 may have had a benign breast mass, possibly a fibroadenoma, with her pain resulting from significant fibrocystic disease. Most breast cancers are not tender, although some women will have atypical breast pain associated with the cancer. Today, she would start with an ultrasound to learn more about the nature of her mass, with addition of mammography as needed. She would have an 11-gauge core needle biopsy for definitive pathologic diagnosis. If her lesion was benign, it might be left in place, or removed for cosmetic purposes or comfort as an outpatient procedure under local anesthesia. Hormonally related fibrocystic breast pain is often managed with non-steroidal anti-inflammatory agents.

Dr. Barbara Lynn Smith

Patient #46: Sprained knee

THEN

A male, age 23, was admitted September 3, 1822, with a sprained knee. He slipped while trying to put grain into the stable four or five days earlier and he struck his knee against the wheel barrel handle. At first he experienced no discomfort. He continued to work during the day until three days later when pain and swelling occurred to a considerable degree and he found himself unable to walk. Within a few days the pain was relieved by local applications with spirits and poultices. Upon admission to the hospital, the swelling was very great around the ankle, with pain and inability to move the knee joint. Then the swelling subsided over his knee, and the pain in the ankle was gone. He still had great stiffness of the limb and inability to bend the leg. He was given alcohol. He continued to do well. On September 9, he was much improved. Blisters grew. On September 24, the ankle vesicated and continued the liniments to the knee. On September 27, he got sulfa and liniment to the knee. On October 5, he felt very well except for a little soreness walking.

NOW

Today this patient would be seen by his local primary care physician, who would refer him directly to an orthopedic surgeon or have him evaluated in the emergency room. This patient would have an orthopedist specialist available to see him 24 hours a day and would likely need a tap of his knee. He would also potentially have an MRI to evaluate the integrity of his ligaments if there was a suspicion of disruption based on physical exam. A culture would be sent of the fluid, and antibiotic therapy would ensue, tailored to the culture information. He would receive physical therapy to improve the function of his knee and accelerate his time to returning to full function on the farm.

Orthopedic Service

Patient #47: Parotid tumor

THEN

A 34-year-old man entered "the house" on October 10, 1822 for the purpose of having a tumor extracted from the back part of the face, over the body of the lower jaw. The history of the tumor was as follows: about twelve years earlier, it was first observed and was almost the size of the end on one's thumb and perfectly indolent. Its growth was very slow in the beginning and on one examination, a hard body was present, situated over the inferior portion of the body of the lower jaw and extending backwards, about one inch, with the posterior portion crossing the temporal artery and lying behind and under the lobe of the right ear. Its projection from the face was more than a half inch. It was well circumscribed and movable under the skin with no great difficulty. There was no pulsation palpable. Its increase in the previous eighteen months had been great. It had been appreciating for ten years and within a few days, had been slightly painful, extending downward about one inch from the zygomatic process, on the temporal bone from which it seemed to be caused. On October 12, preparation was made for an operation with scalpels, warm water, lint, adhesive straps, ligatures, and band aids.

The patient was placed in a chair and the neck uncovered and incision over the middle of the tumor laid open the teguments and exposed the tumor. A careful dissection was now made around the tumor, occasionally cleaning a small quantity of blood which came from the small vessels about the tumor. The full extent of the tumor was soon ascertained and rapid dissection completed; the operation had loss of only a few ounces of blood. Three small arteries branched to the temporal. The wound was now well opened, three small arteries were cured by ligature and adhesive straps were applied, bringing the wound in contact over a compress of lint and linen and then secured by a double-headed roller. Considerable pressure was made by the dressings to prevent hemorrhage that might occur from vessels too small for a ligature. The weight of the tumor was about an ounce. The operation was performed and the wound dressed in about 45 minutes. The doctor commented on the patient's fortitude.

During the operation, the patient occasionally took wine and water to drink. There was no hemorrhage from the wound. The patient was irritable in bed and he took a little coffee. On October 13 he had a tolerable night and slept some without medicine. He complained of slight soreness in the wound extending over the face. In the afternoon, the pain in the face somewhat increased with some swelling. Bowels moved the day before. Patient was lucid. On October 14, he had no operations or medicine; swelling had increased and extended over mouth and right eye. On October 15, soreness and swelling was less. The bandage became loose on October 15. Lint and simple ointment with compressing bandages were applied. On October 16, the wound looked well, pain and swelling diminishing. On October 17, all nearly gone. Now slept late at night. After one dose no pain at all, slept well. On October 18, tongue was white, appetite was tolerable, pulse was natural. He wished to go to Hingham. He was given instructions for the anterior part of the wound, under the skin, pressure over this part. It was dressed lightly with lint and the patient was discharged "cured."

NOW

On October 10, 2011, a 34-year-old man presented to the Division of Plastic Surgery for evaluation of a tumor over the angle of the mandible. The patient first noticed a lump the size of "the end of his thumb" just below his ear about twelve years earlier. It remained the same size for many years but over the past eighteen months it appeared to grow rapidly. In the past few days the mass became painful which prompted him to call his PCP for a referral. The patient denied any other symptoms.

On examination, a firm mass was palpated between the angle of the mandible and the ear. Its projection from the face was more than a half inch. It was well circumscribed and mobile. There was no pulsation palpable. It appeared to extend from the tail of the parotid gland. The patient was sent for fine needle aspiration (FNA) which revealed atypical cells consistent with pleopmorphic adenoma. A few days later the patient received a PATA phone screen which cleared him for surgery. Two weeks later the patient was brought to the operating room, where he underwent general anesthesia and a superficial parotidectomy. The surgery was performed via a facelift incision and under facial nerve monitoring. The patient tolerated the procedure well and was admitted for observation. The next morning, his face was examined. There was no evidence of hematoma and facial nerve function was intact. The drain was removed. The patient was discharged to home with oral pain medications. His post-operative course was uneventful.

Dr. William G. Austen, Jr.

Patient #48: Hemorrhoids

THEN

Patient is a 54-year-old male from Salem, admitted on October 12, 1822. He'd had external hemorrhoids for the last 25 years which were painful and caused by constipation and frequent exposure to bad weather as the master of a merchant ship. There were two protruding "tumors" outside the grip of the sphincter muscle. The largest was about the size of a large pea. The two hemorrhoids were free from inflammation. The operation used lint, sponges, water, olive oil, a tea bandage with ligatures, and instruments.

The procedure was done with the patient in a sitting position, his feet on the floor; the little finger of the surgeon's left hand was oiled and passed into the rectum to bring down the whole hemorrhoid. A hook of some sort was used to pass through the largest tumor and drawn downwards to externalize it to its full extent. A single incision removed the tumor with a scalpel with a loss of a very small quantity of blood. The second tumor was treated in the same way. A little lint was used for coagulation and passed with a probe into the passage. The parts were then dressed with lint. Patient was given wine and water for faintness and pain, continued through the day, but was not urgent. No hemorrhage of consequence occurred. By 9 p.m. he complained of the inability to urinate. Fomentations were placed in the gastric region and he voided a large amount of urine. His diet was tea and gruel. On October 13, he had a tolerable night, no hemorrhage, had the ability to urinate, and no bowel evacuation. Fomentations were placed again and he received sulfa soda aqua and another medication, which he was instructed to take immediately with warm water, two to three times a day. On October 15, there was no bleeding, he had a good night, was very comfortable. On October 16, he had a bowel evacuation without medications and felt very well. He was given bread. On October 17, his bowels were opened and he had discharge but not much bleeding. He was given olive oil to ease passage of stool and on October 19 he was discharged "cured."

NOW

This patient had a limited hemorrhoidectomy. Today, of course, such an operation would be done under anesthesia. I suspect it would have been very painful to pass a hook through the hemorrhoids, pull them out of the anal canal, and amputate them with a scalpel; it's no wonder he needed wine and water for "faintness and pain." Also, modern anorectal surgery is usually done in the prone jackknife position, which gives optimum exposure. It's difficult to imagine how the surgeon could see anything with the patient sitting upright (was the surgeon working under a fenestrated chair?). Post-operative urinary retention is a complication that is still seen today after anorectal surgery, but it is usually treated with insertion of a bladder catheter rather than warm packs to the abdomen (maybe "fomentations" should be tried in the future?). Post-operative pain would now be treated with narcotic medications or nonsteroidal anti-inflammatory drugs, rather than alcohol. The patient spent a week in the hospital, but today this surgery is almost always done as an outpatient.

Dr. Paul C. Shellito

Patient #49: Anal fistula

THEN

A 17-year-old man from Lancaster came into the hospital on October 21, 1822. About four years earlier, he had developed a small hard swelling of considerable soreness at a distance of about 1 ¾ inches from the left side of his anus. On examination, two fistulous ulcers appeared about one half inch apart. On passing the probe into each, communication was discovered about 1 ¼ inches from the skin. The probe passed without difficulty. No opening could be found into the second fistulous tract. Discharge was very slight.

The little nodule did not affect his general health; his bowels were normal and his appetite was good. Preparation for his operation was sulfa soda and abstaining from animal food and butter. Tools were: bandage, lint, and compressed sponges, warm water, and curved scissors. The operation was done in the following manner: four fingers of the surgeon's right hand were passed into the rectum and a probe introduced into the ulcers to ascertain their state and extent. No communication was found to exist between the two. Four fingers of the left hand being in the rectum, the skin and a portion of the cellulous substance were divided, a space of one half inch toward the anus and through the whole depth of the ulcer. Patient had been held on the bed by five assistants and the surgeon, one on each side, one at the head and superior.

A ligature was placed to assist in extending the parts. An attempt was made to thrust the bistoury (a long, narrow knife) through the rectum but such was the resistance, it was in danger of breaking the instrument and further attempt was useless. The bistoury, being new, was withdrawn and introduced into the wound after removing the blood by sponges and being fixed in a position to guard the parts above. The bistoury was again introduced and the wound opened into the rectum after which the rectum was divided as far as the interval from the sphincter anal muscle on the opposite side of the sphincter muscle. The opposite side of the ulcer was then divided about a quarter of an inch deep and the bistoury withdrawn. The sphincter muscle being undivided, a pair of curved scissors was opened and one blade into the rectum and the other blade in the wound and the sphincter was divided by a simple cut which completed the operation. After division of the sphincter, a copious evacuation of feces followed. During the operation, the patient complained of great pain but bore it very well. The wound was thoroughly sponged and the hemorrhage was soon checked and a quantity of lint was introduced sufficient to fill the vessels. Lint compresses and a T-binder completed the dressings. As the patient still complained of pain, 30 drops of tincture of opium were administered.

The patient had a light diet of gruel and tea, a half pint of the former every 3 hours, day and night. He was comfortable on October 24, the pain in the bowels improved somewhat. He took sulfa soda and water on October 25. He did not have much pain. On October 26, the pain in the bowels was relieved, pulse was 84. On October 27, he had stools, on October 29 his appetite was good, and on October 31 he repeated the oil. On November 1, he was doing well, able to take solid foods. Lint was dipped in solution of alum, grain 6, and water was introduced into the wound daily. He also took cinchona and sulfa soda three times a day. He had stools on November 12, his bowels were regular. His appetite was good on November 15 and he slept well. The wound wasn't healing well at the bottom and terminated in a thimble-like cavity. Sulfa was applied, injected into the rectum, to be continued twice daily. He was discharged "cured."

NOW

This sounds like it was an anal fistula. Anesthesia would be used for such an operation today; it must have been painful to insert four fingers into the rectum, and thrust with a bistoury with enough force to risk breaking the instrument! The fistula was apparently completely cut open, which is often done today for superficial fistulas (involving minimal sphincter muscle). This, however, sounded like a rather deep fistula (hence the "copious evacuation of feces" after fistulotomy). Today, such a deep fistula probably would be treated with a muscle sparing technique, such as insertion of a collagen plug into the fistula, ligation of the intersphincteric fistula track, or perhaps staged fistulotomy with a seton. Frequent warm water baths would be used for post-operative wound care today, instead of caustic compounds such as alum or sulfa. Narcotics would be given for pain relief, as this patient received. Nevertheless, cinchona and sulfa soda (presumably purgatives) have been abandoned in favor of fiber powder to keep stools soft during the healing period.

Dr. Paul C. Shellito

Patient #50: Migrating joint pain

THEN

A 21-year-old male admitted October 22, 1822. He had a scrofulous appearance and suffered sudden chills. At one point in time he was unable to raise his forearm to his head, yet later that day was able to dance without symptoms. Then he had pain over the os femoris. Half an hour later couldn't even put his foot to the ground; his knee was red and swollen. He suffered ten days of increased pain of the knee and emaciation of his lower limb. On October 25, he had poultice applied to the knee every four hours made of rye meek and milk. On November 1, he had pain in the knee due to the application of caustic medication. On November 4, he had pain to his lower shoulder and breath, pain moved to the tibia. Four leeches were applied. On December 10 his pain was very minimal. On February 19 he was discharged.

NOW

Minor orthopedic condition

Patient #51: Cataracts

THEN

A 75-year-old woman came into the hospital on November 2, 1822 with cataracts in both eyes. She thought she got them from the exertion of sewing and by exposure from sleeping near an open window. She was totally unable to discern common objects. The left eye was almost totally blind, however, the right eye remained unaffected during the progress of the disease in the left and so, sometime in April, there was no remarkable symptom which indicated that she was going to have anything in either eye except for an almost intolerable itching of the eyelids, constant rubbing or scratching and generally during the daytime. Exposures to light were not painful. Her medication included r.linci sulphat grain 6 and water applied two to three times a day via an eye cone laid over her eyes. She also got sulfa soda and a normal diet. Belladonna occasioned some smarting of the eye; cataracts appeared to be yellowish color. Patient could not distinguish much light or any objects perfectly. The eyes were washed, pupils dilated with belladonna, and the eyes looked less inflamed.

She was prepared for the operation. The patient was to be in a cool and tranquil state with moderate use of purgatives. She would be purged the day before the operation but not on the same day. The susceptibility of the eyes to belladonna would be determined. If previous to the operation, the belladonna applied at such time before as to cause the greatest dilation.

The patient's seat, as well as the surgeon's seat, would be adjusted as much as possible before the patient was called on. To be used: soft bandages, soft compresses of linen, bits of lint, wine and water, scoffers, cataract knife, block for assistant, and warm and cold water. For the operation, the assistant tried the speculum, adjusted the patient's head and fixed the arm position which he and the patient were to preserve. An assistant on each side held the patient and a third held the instruments. After he took notice of the direction in which the light is to reflect from the eye, placement of where the needle is to enter, a line behind the cornea, the steadiness of the patient's eye, and controlling it with four fingers of the left hand. He entered the needle perpendicularly to the center of the eye and afterwards, brought the needle forward, toward the pupil. Next came the point near the internal margin of the iris and pressing the needle back, determined the movement of the cataract, and whether it were hard or soft. If hard, he would incline the point backwards, upward and downward until the cataract disappears. If the cataract were soft, he'd turn the edge of the needle backwards and carrying it in various directions, dividing the lens into many pieces. Then with the point of the needle, he shoves as many pieces possible into the interior chamber.

The right eye should be subsequently operated on, the sheets should be changed and everything done as just directed. After treatment, the patient should be carried to a dark room, eyes previously covered with bandages and treated for pain. Some blood in the eyes, eyes bathed in warm water, perhaps leeches applied if necessary. On November 22, her appetite was good. Right and left eyes were treated with belladonna, the left eye dilated but the right eye did not. Belladonna made the eyes feel more comfortable. On November 23, the patient being seated, the speculum was applied to the left eye. The needle was passed into the globe of the eye, about a line distance from the internal edge of the cornea, and the cataract was discovered to be capsular and soft, and was immediately cut open into numerous pieces. There was no escape of the agueas.

The cataract being sufficiently divided and broken pieces were lodged in the interior chamber of the agueas and the instruments were withdrawn. The chairs and stools being reversed, the same operation was performed on the right eye but passing the needle through the coats of the eye and examining the cataract, it was found to be more solid and resistant to the instrumentation. It was divided into many; the left was more solid and more resistant to the instrument than that of the right. The operation was finished without pain to the patient, nothing more than a few bits of linen and soft compresses and bandages with dressings. Patient was removed to her bed, given belladonna every four hours to the right eye and every twelve hours to the left. Leeches were applied to the temples. On November 24, a little pain in the eye. On December 2, had nausea and vomiting since November 25 without apparent cause. Leeches were applied to the temples. They applied as many leeches as would hold to take down the swelling. The eye was washed every four hours in warm water. The edge of the right eye was covered with fragments of cataract. They repeated the blisters and leeches. Cataract in the right eye was thinner, applied belladonna once a day, thoroughly. On December 20, the left eye had inflammation and the patient could not see very well. The right cataract was dissolving very well. Blistering was repeated as was application of belladonna.

On January 4, cataracts were dissolving rapidly. On January 8, she saw light, more from the left eye than the right. The left pupil was perfectly clear. On February 7, the left eye was nearly clean and could distinguish colors. Both eyes continued to improve. On March 1, the opacity of the left eye diminished, the right eye had blistered back to the neck. The cataract in the right eye was very slow in absorbing as was left eye. She could see pretty well with left eye. The operation was to be repeated in right eye on March 16. Preparation was the same as before. Patient had been gently purged.

Placed in the chair facing the window, with assistants on each side, the speculum was applied. A slender needle was passed into the globe of the eye and the residual placed about 1 ½ lines distance from the edge of the cornea. Its point was dissected towards the pupil. On examining the cataract, it was found to be exceedingly firm and resistant. Every attempt failed to divide the cataract and the instrument was withdrawn after a few minutes. A sharper needle was introduced and the cataract confirmed by a firm pressure and somewhat broken into, by rotating instrument on the body of the lens. The cataract being fixed at the bottom of the eye, the instrument was withdrawn. On the afternoon the procedure was concluded, some light was visible. Dressing consisted of a bit of linen cloth laid lightly over the eye and secured by a bandage. Compresses and leeches were applied. On March 17, the patient noted that she had little pain since the operation and slept entirely during the night. Extract of belladonna and aqua and water were placed over the eyes and blisters created behind the ear. On March 15, leeches were still being applied. The eye had been quite painful. Finally, on April 25, she thought she could see with the eye, enough to walk around during the day. Doctors kept putting blisters behind the ear. On May 24, there was no pain in the eye. Inflammation had increased. Two pills every other night. On May 26, she was eating. By the end of May, she has no inflammation of the eyes and by the use of cataract glasses, she can see to walk. The patient was discharged July 22, 1823, much improved.

NOW

This 75-year-old patient had visually significant cataracts ("*She was totally unable to discern common objects,*" "*Patient cannot distinguish much light or any objects perfectly*") in both eyes. Cataract is opacity of the normally transparent crystalline lens that degrades the optical quality of the lens and leads to deterioration of vision. Cataract still remains the leading cause of treatable blindness worldwide. Depending on the location of opacity within the lens, cataracts are classified as nuclear, cortical, or subcapsular. Nevertheless, they are all treated similarly, that is with removal of the opacified lens and implantation of a new, clear prosthetic one.

Since antiquity, and until the mid-20th century, luxation of the lens back into the posterior chamber as a whole or in pieces was the main surgical technique used – this was termed "couching" – and is exactly what this patient underwent in 1822. The "hard" cataract in her left eye was depressed as a whole into the posterior chamber ("*If hard, he [the physician] inclined the point [of the needle] backwards, upward and downward until the cataract disappears.*"). The "softer" cataract in her right eye was broken into pieces that were then pushed back into the posterior chamber ("*If the cataract is soft, he turns the edge of the needle backwards and carrying it in various directions, divides the lens into many pieces, then with the point of the needle he shoves as many pieces possible into the interior chamber.*").

Today, this patient would also undergo cataract surgery but with a different surgical technique and certainly before her vision deteriorated to light perception. Cataract extraction is indicated today when visual function no longer meets the patient's needs and cataract surgery offers a reasonable likelihood for improvement in both visual and physical function. Improvement of visual function includes better optically corrected visual acuity, increased ability to read or do near work, reduced glare, improved depth perception, color vision, and binocular vision. Improvement in physical function includes increased mobility, increased ability to perform daily activities, and fall prevention.

The initial work-up today consists of a comprehensive ophthalmic evaluation in order to confirm that a cataract causes the visual symptoms described by the patient and to exclude other ocular or systemic pathologies that could account for the patient's visual impairment or that could affect surgical management, post-operative treatment, and risk for complications. If eyeglasses or visual aids are found to correct the patient's refractive error and provide vision that meets the patient's needs, cataract surgery is deferred.

The surgical method employed today in cataract surgery consists of extraction of the cataract (not luxation into the posterior chamber) and subsequent implantation of an intraocular lens to correct the aphakia (and the 20-diopter refractive error that would result from it). Like in the 1800s, the most popular technique today consists of dividing the cataract into pieces, which are then emulsified with ultrasound and aspirated. Most cataract surgeons use local anesthesia, i.e., retrobulbar, periocular, or sub-Tenon's injections combined with topical and intracameral (injected into the anterior chamber) anesthesia. A five percent solution of providone iodine in the conjunctival cul de sac is used at the beginning of the operation to prevent infection.

A few days after her cataract operation, pieces of the fragmented lens were seen in the anterior chamber of this patient's right eye. The resulting inflammation in the anterior chamber likely led to a rise in intraocular pressure and the symptoms of pain, nausea, and vomiting that the patient experienced. Thus, a re-operation in that eye was performed in order to push more lens fragments back into the posterior chamber. Inflammation quieted down and the patient was discharged home several weeks afterwards able to see *"enough to walk around during the day"* with cataract glasses – thick convex lenses that aimed to replace as much refractive power as possible of the removed natural lens. Since the natural lens has about 20 diopters of refractive power and glasses could only correct for five to seven diopters, patients were only able to see crude object shapes after cataract surgery though the vision was still better than prior to the operation.

A rise in intraocular pressure due to blockage of the aqueous drainage system of the eye by retained lens fragments and/or due to an inflammatory response to them is rarely observed today after cataract surgery. The most feared complications in the perioperative period that can cause permanent visual decline include suprachoroidal hemorrhage, cystoid macular edema (CME), retinal detachment, corneal edema, intraocular lens dislocation, and infectious endophthalmitis. The most common cause for a rise in intraocular pressure post-operatively is retained viscoelastic material, which is used during surgery to protect the corneal endothelium. Thus, early follow-up and ophthalmic examination within 24 to 48 hours from the time of surgery is crucial for recognizing and managing complications and is exercised universally by anterior segment practitioners. Choice and frequency of topical antibiotics (for prevention of endophthalmitis), corticosteroids (to reduce intraocular inflammation after the surgical manipulation), and NSAIDs (for prevention of macular edema) post-operatively varies among cataract surgeons. Repeat refraction is usually performed one month after the surgery and new optical correction (if needed) is provided to the patient.

In contrast to the 1800s, cataract surgery is an outpatient procedure for over 95 percent of patients operated on today. Systemic comorbidities, surgical complications, or poor compliance with post-operative care are some of the indications for hospitalization after cataract surgery today. Modern cataract surgery techniques have been refined to the point that 20/20 results are now routine and in fact expected by most patients. New intraocular lens choices can minimize the need for glasses after surgery such that patients can now return to doing most of their activities with exceptional confidence and convenience. Many recent studies have demonstrated the long-term medical benefits of cataract surgery for the elderly, including a reduced risk of falls, which can lead to hip fracture.

Credit should again be paid to early eye surgeons for their ability to perform procedures such as these without microscopes, proper instruments, anesthesia, or even modern lighting. Great powers of reassurance must also have been at their disposal considering patients such as this one were admitted for over six months at a time.

Dr. James Chodosh
Dr. Matthew F. Gardiner
Dr. Joan W. Miller
Dr. Sotiria Palioura
Dr. Athanasios Papakostas

Patient #52: Infected puncture wound of foot

THEN

A male was admitted on November 7, 1822, having suffered a wound on his left foot from the fall of a pen knife into the tendon of the tibialis over the phalangal joint of the great toe. He had lameness when attempting to walk and an abscess had formed under the skin. Treatment included massaging the skin vertically, filling the wound with lint, and dressing it with a bread and milk poultice. On November 15, the wound was almost well. The next day he had chills, fever, and headache, which lasted one day. On November 22, he was discharged, with the wound fully healed.

NOW

A male patient presenting with a wound such as this, with inflammation and an abscess, would have the abscess drained and dressed. The patient would be given antibiotics and sent home.

Dr. George Velmahos

Patient #53: Open fracture of humerus

THEN

A 26-year-old male was admitted November 25, 1822. In September of that year, he had been involved in a rock blast and was struck in shoulder with a rock weighing six pounds. He had a fracture of the humerus just below the neck. The fracture was reduced and he was immobile for 35 days. He had a fistulized abscess on top of the internal edge of the deltoid. The fistula was four inches and extended to the bone. Treatment consisted of rubbing with oil and turpentine. He underwent an operation on November 28. A poorly united fracture was found. The probe was passed to the bone. An incision was made to the edge of the deltoid two inches. The sharp edge of bone was removed by chisel. The wound was filled with lint. The patient fainted and was given wine and water. Compresses were applied. On November 30 the shoulder was painful and swollen and he was given bread and poultice. Lint was removed. On December 8 healthy granulations were given. In the middle of treatment the patient developed gonorrhea and the prepuce was divided for a length of one half inch. On February 28 a probe passed the bone was found to be smooth. On March 11 the arm wound was closed with granulation tissue and he was able to raise his arm to 90 degrees. On March 21 he was discharged "cured."

NOW (See Patient #35)

Patient #54: Blunt trauma to finger

THEN

A male entered the hospital on November 25, 1822 with a painful left arm and forearm as a consequence of bruising the middle finger of the same hand with a hammer on the November 16. Since the accident, although he had inflammation and swelling that had been subdued by poultice, he still complained of very severe soreness. His soreness extended from the wrist to the shoulder and coursed the circumflex and spinal nerve and the nerve that passes over the internal condoyle of the os-humerus. He had formulations applied whole and then blister was applied the next day above and below the elbow. He took some opium. On November 28, the pain was severe, especially at night. He was given medication to take until vomiting. He took extra laxative to open the bowels every other day with magnesium sulfate, and to the most painful part a poultice of rye meal with milk, a half ounce of white poppy heads every four hours as long as he could bear a really hot pack. Hot packs were continued until he had no pain and he wished to be discharged. In summary, he had swelling which was cured by hot packs.

NOW

Minor orthopedic condition

Patient #55: Tumor of foot

THEN

A male, age 22, entered the hospital November 26, 1822 with a small, well circumscribed tumor over metatarsal bones and the second and third toes. His left foot was somewhat painful when walking. His general health was good. An operation was done. The instruments were blunt and sharp scalpels, hooks, sponges, hot and cold water, wine and water, adhesive, straps, and roller. He was secured on the table by two or three assistants and an incision was made through the skin directly over the center of the tumor. The skin was then carefully resected on each side of the tissue and the whole removed without loss of any considerable quantity of blood. The operation concluded without much pain to the patient. The wound was well sponged and dressed with adhesive tape. The weight of the tumor was one ounce. The wound was dressed daily with adhesive straps and linen with magnesium sulfate.

On December 4, he had some pain in the foot and applied a poultice of bread and milk and was given cathartics. There was some inflammation around the wound as a consequence of his walking around and removing the straps of his own accord. Straps were reapplied; the foot was free from pain and soreness on December 8. He was discharged on December 17.

NOW

No commentary

Patient #56: Rib abscess

THEN

A 20-year-old male entered the hospital November 29, 1822. A sudden cold left him with hardness and pain over the tenth rib. By advice of the surgeon, poultice was applied to the body part and continued for three weeks, after which an abscess formed, which burst and discharged whitish matter and continued to do so until admission to the hospital. He felt weak and his appetite was not strong. He felt weak while exercising. His pulse was 86 and firm. He had a slight tremor to the hands. When he returned last week, he had a dull pain with soreness and inability to bend arm. The swelling, he said, was removed by bandaging. As of November 29, he had a slight dry cough. His respiration was not much affected, but he had a slight pain in the lower left side of his chest and around the anterior extremities. He was given magnesium sulfate and menthol and pulverized cinchona, three times per day and caustic potash applied to the arms under the elbows. In summary, he had soreness on the sides, less painful. No discharge. The abscess that had formed was dressed daily.

On February 5 he had an operation. The instruments were scalpel, probe, ligature, crooked needle arm ligature, sharp chisel for removing dead portions of bone, sponge, lint, hot and cold water, and a wide bandage for the chest. The patient was prepared for the operation with 80 drops of opium given pre-operatively. He was secured on the table by assistants. The probe was passed into ulcer to ascertain the extent of the disease in the ribs, this being done with an incision of the rib made through the skin and the fibers of the radius major anticus muscle down to the bone and extended about 2 1/2 inches. The ulcer overlied the middle part of the incision. The bone was fairly exposed. A small portion of the inferior margin of the seventh inferior and superior eighth rib, and a few small fragments adhering to them, were removed. These fragments were disengaged by careful resection and the bones were rendered quite smooth by means of a chisel. The edges of the wound were completely removed and then the new edges were drawn together with one stitch in the middle and the rest was retained by adhesive and a simple compress of wide roller, which finished the dressing. "The operation was born with a good deal of fortitude," noted the surgeon.

Post-operative directions were to take arrowroot, light nourishment, and to keep perfectly quiet. He slept through the night. His pulse was 104, tongue thinly coated. He got magnesium sulfate, but the medicine did not work. On February 8 he had had no stools since the operation. On February 12, the wound appeared healed. Bowels were well, appetite was good, and patient was discharged "cured."

NOW

This patient would be diagnosed with an abscess and started on antibiotics if there were associated cellulitis. Arrangements would be made for an incision and drainage of the abscess cavity early. A bedside incision and drainage of the abscess would be attempted with the assistance of local anesthesia. If the abscess cavity were larger or more complex, then arrangements would be made to perform a more aggressive drainage and debridement under general anesthesia. The post-operative course would likely be similar to the THEN scenario. If the wound were large, a vacuum-assisted dressing would be applied to accelerate wound closure and debridement.

Surgical Service

Patient #57: Uveitis

THEN

A 22-year-old woman entered the hospital on December 2, 1822. Over the previous two years she had had a steady fixed pain in both eyes which increased upon exposure to light and motion. At the same time, she occasionally felt dizzy and light-headed, with nausea and no appetite. Over the first four months, she got much advice and medicine without any relief. Her general health was good except for occasionally a disagreeable sensation. Her eyes presented no unnatural appearance. Her iris contracted readily upon exposure to light but not more than is usual, and dilated again when removed from a stimulus.

During the last six weeks before admittance, her eyes had been inflamed. Once at the hospital, she was given medication, bread with milk and tea, and did not eat animal foods. On December 11, she complained of pain in her head. She was venesected with no relief from the bleeding. She also complained of pain over the lower side of her back, which had been troublesome for some months. She had pain in the right hypochondrium, at times severe. Her pulse was 75. Blisters were applied to the lower part of the back. On December 13, the pain had continued through the night, was severe, and she had no appetite. She had multiple, various drops put in the eyes, even extract of belladonna. She had colchicum for a while and then was discontinued. She was venesected without relief. Blisters were applied to the back. On January 16, plaster was applied to the head, three quarters of an inch by one and a half inches. On January 27, she was venesected with mild relief. On February 2, she was bled with some relief. They treated her with blisters and bleeding. She had conjunctivitis bilaterally. Blisters to the right eyelid healed. She felt like she had something in her eye. On February 15, the pain in the right eye was severe. On the February 16, reduced pain in the eye continued with double vision. Poultice continued to be put on her back and the blister in her eye continued to be painful. On March 5, with a moisturizer infused with chamomile flowers and continued venesection performed, her eyes were a little better. Despite ipecac and magnesium sulfate, eyes and head remained painful. On April 6, she was discharged.

NOW

This patient had pain in her eyes upon exposure to light and motion, right upper quadrant abdominal pain ("hypochondrium"), low back pain, sluggish response of the pupils to light (without an afferent pupillary defect), and inflammation in her eyes. All these findings suggest the presence of an inflammatory eye disease, most likely uveitis. She had probably experienced multiple attacks in the past which created posterior synechiae (thus the sluggish pupillary responses), and this time presented with an acute episode. The other systemic findings suggest an underlying disorder responsible for the inflammatory eye disease. The right upper quadrant pain could have been pleuritic and the low back pain could have been secondary to inflammatory joint disease (ankylosing spondylitis) of the spine.

Based on the above findings we can formulate a differential diagnosis which includes HLA-B27 associated uveitis, reactive arthritis, sarcoidosis, tuberculosis, syphilis, and systemic lupus erythematosus. An appropriate work-up would include a complete blood count with differential, erythrocyte sedimentation rate and serum C-reactive protein, placement of a PPD, chest X-ray, ACE, lysozyme, RPR/FTA, VDRL, ANA, p-ANCA, c-ANCA and HLA-B27. Treatment initially would include topical prednisolone every hour along with a dilating agent. Later in the course of treatment, an immunomodulatory systemic agent could be added based the lab findings. [See also Patient #71.]

This patient endured all of the therapeutic modalities popular in the early 1800s, including bloodletting, blister creation, and purging via ipecac without relief of her symptoms. Uveitis can sometimes resolve spontaneously making it difficult for early practitioners to know with confidence if any treatment was truly responsible for recovery. The one therapy she received that is still a mainstay of treatment for uveitis is pupil dilation (now with topical atropine, but then with belladonna) to prevent intraocular adhesions between the iris and the lens. The unavailability of immunosuppression to her physicians made it unlikely that her treatment would have been successful.

Dr. James Chodosh
Dr. Matthew F. Gardiner
Dr. Joan W. Miller
Dr. Sotiria Palioura
Dr. Athanasios Papakostas

Patient #58: Soft tissue injury

THEN

The male patient was admitted on December 16, 1822 as a result of falling off a two-story house while attempting to extinguish a fire. He was stunned by the fall, had several wounds about his face, contusions about the teeth, and considerable lacerations about the forehead. His hands and wrists were swollen and very painful. There was no fracture of these bones. His face was generally swollen. His wrists and hands were moistened with equal parts vinegar and water and simple dressings were applied to the wounds. He got magnesium sulfate. On December 17, the hands and wrists were less swollen than the previous day but still painful. Bread with milk poultice was used. Wounds were washed with soap and water and wrapped with a dressing of diluted ammonia muriatic alcohol. No other notes were made. On the December 25, nine days after his admission, he was discharged "cured."

NOW (See also Patients #12, 14, 39, 55, 58, 87, 95)

Nearly 200 years ago, physicians at the Massachusetts General Hospital were equipped with a variety of tools for the care of both acute and chronic soft-tissue damage. These bear remarkable similarities as well as striking differences to those used in 21st century medicine.

Striking Similarities in Wound Care Management – Then and Now

Two key tenets in the modern treatment of soft-tissue injury are observed in many of the cases first described in the history of MGH. These include: 1) the use of occlusive dressings for wound coverage; and 2) vigorous wound care with manual debridement and cleansing agents.

Patients admitted to MGH with traumatic, surgical, or chronic wounds all underwent daily dressing changes with attention drawn to the care of the wound at least once a day. For instance, Patient #58 suffered a fall from a two-story house and sustained multiple traumatic lacerations to his face and hands. While in the hospital, his wounds were washed daily with soap and water, and wrapped daily with occlusive dressings. Other patients, including Patients #12, #14, #39, #55, #87 and #95, were all described as having open wounds, which were dressed daily with a variety of methods including cloth, cotton pledgets, linens, lint, simple ointments, plasters, and adhesive straps. In modern day medicine, research has led to the treatment of wounds with a variety of dressings, ointments, and adhesive materials, based on their specific etiology and characteristics. To this day, daily rounds on a surgical service at MGH include wound maintenance and dressing changes. Surgical wounds are typically dressed with dry, sterile gauze made with a cotton base (not unlike the cotton-based cloth, pledgets, and linens of 200 years ago). Current dry dressings for clean, surgical wounds differ in their sterility and composition from those of the past, but the principles remain quite similar.

Chronic ulcerative wounds in modern medicine ideally are treated with a wound dressing that "removes excess exudate, maintains a moist environment, protects against contaminants, causes no trauma on removal, leaves no debris in the wound bed, relieves pain, provides thermal insulation, and induces no allergic reactions."[1] Examples of such dressings include films (thin, semi-occlusive membranes that allow an exchange of oxygen and water vapor between the wound and the environment), hydrogels (water-based products used to maintain a moist environment), hydrocolloids (absorptive substances), alginates (absorbent fibrous dressings), collagens (matrix-forming dressings), and foams (absorbent semi-occlusive dressings).[1] Modern day wound care emphasizes the presence of moisture and occlusion as an aid in the promotion of chronic wound healing via increasing re-epitheliazation, collagen synthesis, and promotion of a low-oxygen environment with antibacterial properties.[2,3] In the cases described above, pioneer physicians at MGH utilized ointments, poultices, and plasters on wounds with positive effects due to their inherent occlusive and moisture-inducing properties. It was not until at least 150 years later that science would fully characterize the nature of these effects!

The second principle that bears a similarity to modern medicine is the debridement and maintenance of a clean wound. In earlier years, wounds were often treated at MGH by washing and debridement with soap and water, as well as chemical debridement via a caustic solution (acid liniment, vinegar, magnesium sulfate paste, acid alcohol, etc.). While some of the antibiotic properties of these chemicals have not been borne out in the literature, the importance of a clean wound has not changed in the past 200 years. Besides manual debridement of wounds, modern medicine uses many chemicals (not unlike those used in the past) for their antimicrobial properties and to promote wound auto-debridement.[1] Hydrogen peroxide, iodine, silver-releasing agents, and Dakin's solution are currently used for their "caustic" properties in wound care with excellent antimicrobial and cleansing efficacy. One difference, however, between soft-tissue wound care then and now is that caustic solutions are currently only applied in the *initial* phases of care, as opposed to the constant application of caustic solutions during all phases of wound care 200 years ago. Applying caustic solutions to a wound while in its "healing phase" will limit the progress of fibroblastic proliferation, and cause healing to slow substantially.

Advances in Soft-Tissue Injury Care in the 21st Century

Despite these similarities in care practices between MGH then and now, two particular cases from the MGH records highlight significant differences in the way that soft-tissue injury management has evolved over the past 200 years. Patient #87 details a 17-year-old admitted in June of 1823 with "rheumatic pain." Incidental note was made of a skin avulsion to the front of his shin "about the size of a 25-cent piece." This, according to the rest of the case, had become a significant problem for him. Skin on the anterior portion of the leg (pretibial region) that is lacerated or avulsed can be prone to a multitude of complications given the lack of underlying soft-tissue coverage and the prominence of the anterior ridge of the tibia. This is particularly worrisome in full thickness tissue loss associated with fractures of the tibia or in avulsion injuries for patients with poor quality skin.

Physicians who treated Patient #87 described covering the pretibial wound with ointments and daily wound dressings with limited success. They then turned to sprinkling "burnt alum" (heated potassium alum, a natural antiseptic agent also used to minimize bleeding) onto the wound three times daily, only to see the ulcer worsen. After two months of hospitalization, his ulcer was improving slightly, and a caustic solution was applied along with daily poultices and ointments. He was finally discharged after a four-month hospitalization, "relieved," although no mention was made as to the status of his pretibial wound.

The management of pretibial wounds in modern medicine is complex, and depending on the extent of the wound, a variety of treatment options exist. For acute partial thickness skin avulsions without exposed bone, patients are typically treated with petrolatum and antibiotic-impregnated gauze applied to the wound along with daily dressing changes. Patients can also have skin grafts applied to any areas of exposed muscle, but not bone, for larger wounds with significant de-epithelialization.[4] For full thickness skin loss with exposed bone, patients can undergo a rotational flap, where one of the posterior calf muscles is transposed to cover the exposed bone, and that is subsequently skin grafted. Additionally, healthy patients can also undergo a perforator-based advancement flap as well as a "free flap."[5] Both of these options involve the transposition of healthy skin (and its vascular pedicle providing blood supply) to the diseased area by either utilizing rotation/advancement of the skin or a free-tissue transfer with a microvascular anastomosis. While the extent of his original wound is unclear, it is likely that Patient #87 could have been treated successfully with one of the above modern therapies.

Finally, Patient #95 deserves particular attention. Three weeks prior to presentation, a 7-year-old female had her right knee crushed between a tree and a large stone, sustaining complex lacerations down to tendons and ligaments. By report, this was closed by an "interrupted suture" on presentation. Unfortunately, the skin started to separate soon after and skin sloughing was present. The wound was then dressed with lint and ointments, and three months later she was discharged with the "wound to the knee…healing well."

Acute massive soft tissue lacerations about the knee can be quite problematic. The young girl underwent primary closure of an open wound three weeks after the injury, and it is no surprise that closure with an "interrupted suture" – whether one or many – had failed. Currently, large acutely sustained wounds are typically managed with early and aggressive irrigation and debridement followed by either primary, secondary, or advanced methods of closure. For grossly contaminated or late-presenting wounds, surgeons may elect to perform multiple-staged wound debridements before definitive closure.

One particular advancement in the field of wound care for large acutely sustained wounds is vacuum-assisted wound closure (VAC). This method exposes a wound bed to negative pressure through a closed system, creating mechanical tension that stimulates cellular proliferation, removes edematous fluid from a wound, and improves blood supply to promote healing.[6-8] It can be used in conjunction with multiple surgical debridements, and probably would have been utilized on this child prior to definitive closure. Also, proper debridement of dead or infected skin may have precluded primary closure of the entire wound, and secondary intention or more advanced closure methods (similar to those described above) may have been utilized at MGH today. At discharge, it is not clear whether the wound was entirely healed, and by today's standards, this would not have been considered a success.

Additionally, large wounds around a major joint carry a high risk for stiffness, and range of motion is not mentioned in this patient's final outcome. It can be presumed that a chronic wound around the knee caused this child a great deal of stiffness and had she been treated at MGH today, a large portion of her rehabilitation once the wound was closed would have focused on knee range of motion to prevent contracture.

The aforementioned cases are prime examples of the way in which modern medicine has evolved to treat soft-tissue injuries. While respecting the same fundamental principles, wound care has become a complex field with multidisciplinary care teams working to advance wound healing into the 21st century and beyond.

References

1. Fonder MA, Lazarus GS, Cowan DA, Aronson-Cook B, Kohli AR, Mamelak AJ. Treating the chronic wound: A practical approach to the care of nonhealing wounds and wound care dressings. *J Am Acad Dermatol* 2008;58:185–206.

2. Alvarez OM, Mertz PM, Eaglstein WH. The effect of occlusive dressings on collagen synthesis and re-epithelialization in superficial wounds. *J Surg Res* 1983;35:142–148.

3. Rovee DT. Evolution of wound dressings and their effects on the healing process. *Clin Mater* 1991;8:183–188.

4. Foroughi D, Nouri DK. Grafting without a donor site: an easy approach to pretibial lacerations. *J R Coll Surg Edinb* 1990;35:245–247.

5. Kamath BJ, Joshua TV, Pramod S. Perforator based flap coverage from the anterior and lateral compartment of the leg for medium sized traumatic pretibial soft tissue defects--a simple solution for a complex problem. *J Plast Reconstr Aesthet Surg* 2006;59:515–520.

6. Webb LX. New techniques in wound management: vacuum-assisted wound closure. *J Am Acad Orthop Surg* 2002;10:303–311.

7. Park JJ, Campbell KA, Mercuri JJ, Tejwani NC. Updates in the management of orthopedic soft-tissue injuries associated with lower extremity trauma. *Am J Orthop* 2012;41:E27–35.

8. Wanner MB, Schwarzl F, Strub B, Zaech GA, Pierer G. Vacuum-assisted wound closure for cheaper and more comfortable healing of pressure sores: a prospective study. *Scand J Plast Reconstr Surg Hand Surg* 2003;37:28–33.

Dr. Eric M. Black
Dr. Harry E. Rubash

Patient #59: Soft tissue mass (lipoma)

THEN

A married woman, age 20, was admitted with a large tumor mass situated just above the crest of the ileum on the right side. It had been fourteen years since this was first seen. She did not have any pain, but it grew larger over the first ten years and then, over the last four years the increase had been much more rapid. It extended three inches from the spine posteriorly to about five inches from the groin anteriorly. The substance is soft and flabby but not fluctuating and seems continuous with the common lateral dimension, the widest point about six inches and gradually growing narrower at its extremity. The pressure causes some pain. An operation was going to be performed on January 15. The instruments were blunt and sharp scalpels, and hooks. Other preparations were dressing, adhesives straps, compressors, ligatures, sponges, hot and cold water, wine and water. At 12:00 p.m., she got 40 drops of tincture of opium, given over one half hour, after the patient was placed on the table with her hands gently raised over the pillow. The tumor was examined thoroughly; several assistants secured the patient on her left side, taking hold of the arms, head, and feet. The skin was drawn tight over the tumor and incision through the skin layers was made, exposing the substance of the tumor which consisted of fat. The incision was next enlarged to the extent of four inches in width and one foot in length. The extent of the tumor could not be ascertained. Dissection separated the skin covering swelling each side until the fibers of the external rear muscles on the abdomen were exposed and seen through the fascia covering them. The mass was removed in a few minutes; it weighed two pounds, four ounces. In examining the skin, numerous small growths were found about the wound and also about the body generally, some were removed and found to contain a white, cheesy matter. The wound was put together by adhesive strap over compressive linen. The patient complained of great thirst and exhibited signs of delirium. After being moved to her bed she soon became more tranquil. At night she was stable except for some pain in the abdomen.

On January 19, the wound was flabby and gangrenous. Pulse was 120. The wound was dressed with rice poultice and washed daily with soap. On February 5, the patient could sit up without any inconvenience; the dressing was given various straps and compresses.

On February 23 she complained of pain in the epigastrium with nausea. Pulse was 84. Wounds were well healed except for some granulation tissues. Wound care was documented in Dr. Jackson's book, B, page 202. On February 26 the wound was healthy, and continued to heal without much pain to the stomach and bowels.

On March 26 a caustic was applied to the lips of the wound in order to bring on fusion between them. Wound was healing well. She was discharged on April 5.

NOW

This patient presents with what appears to be a very large subcutaneous lipoma. The management of lipomas of "reasonable" size is excision under local anesthesia. These tumors can usually be "enucleated" as they are encapsulated and seldom invade the soft tissue. Even now, however, for a lesion of this size, which would be uncommon and have to have been essentially "ignored" for years, general anesthesia may have been required due to the size of the incision and dissection necessary. There might be concern that, because of the large dead space left from the excision of this massive a tumor, a subcutaneous fluid collection could develop and lead to a wound complication. Therefore, a closed suction drain might have been considered to approximate the tissues. The skin certainly would have been closed primarily.

The other numerous small growths containing "white, cheesy matter" likely represent sebaceous cysts, which would also be locally excised, again under local anesthesia. It is important that the entire cyst wall be excised with its contents to prevent recurrence.

Dr. Keith Lillemoe

Patient #60: Soft tissue infection

THEN

A seaman was admitted January 3, 1823. At sea he felt poorly and got another seaman to bleed him with a lance. He complained of soreness when admitted, and had numbness of the ring and middle finger. His arm was growing weak. The puncture was at the elbow bend at the brachial artery. The lance was put in about a half inch deep. There was swelling. He had stiffness of the annular ligament. The arm looked emaciated. He complained of knee pain. He was vesicated, and on February 2 he was discharged.

NOW

Today the patient would be evaluated in the emergency room and if there was a suspicion of an abscess in the arm but it was not visible, an ultrasound of the arm may be performed. In addition, if the patient was found to have a skin break from a traumatic injury, tetanus toxoid would be administered. If the patient was found to evidence of septic emboli with infection seeded in his arm as well as his knee, his cardiac valves would be evaluated for evidence of vegetations. The patient may have required long-term antibiotics, which would be administered via a peripherally inserted central catheter (PICC-line) while at home. Other possible medical causes including vitamin deficiencies would also be considered in the differential.

Surgical Service

Patient #61: Contusion of knee

THEN

A male came into the hospital on January 22, 1823 with contusion of the right knee, as a consequence of a wagon having run over those parts that afternoon. Examination discovered nothing important about the bones in the pelvis. He had a contusion on the skin above the right knee. Nothing was broken. He got compresses to the knee and on the January 25 he could walk about the ward and he was discharged.

NOW

Minor condition

Patient #62: Venereal disease

THEN

A male patient admitted to the hospital on January 25 with gonorrhea of three months standing. He had previously had no medical advice. Now, he had a discharge from the urethra of thick yellowish matter. His health was generally good. A good-sized bougie was able to be placed in the bladder. Magnesium sulphate was given. He was discharged without change. He returned on February 2, same as before. The discharge hadn't changed, but his mouth was a little sore. On April 29, a bougie and catheter passed readily without pain or discharge.

He got cinchona to be taken twice a day with milk. He got some exercise and some air, and a warm sea bath. He wished to be discharged and was discharged "cured."

NOW

This dermatology case review of the Massachusetts General Hospital from the 1820s reveals a host of maladies, from syphilis to herpes to eczema. An overwhelming theme in the cases from that time is the genital ulcer. Historically, dermatology and venereology were linked as a single field, as can be seen in early dermatologic texts, such as Josef Jadassohn's *Handbook of Skin and Venereal Diseases*.

Over time, the focus of patients seeking dermatologic care has shifted from the infectious to the neoplastic. According to records from McCall

Anderson[6] in Scotland in 1887, at least 50 percent of 11,000 recorded cases were infections or infestations, in contrast to the modern Scottish Dermatology Clinic where close to 50 percent of visits were for skin cancer.[7] Our own experience at MGH mirrors this trend.

The realm in which we treat patients has also shifted. While at MGH we have a growing inpatient dermatology consult service, which sees a variety of inpatients with leukemia, infection, drug eruptions, and neutrophilic dermatoses, much of present day dermatology care is done on an outpatient basis. Weekly wound care clinics for ulcer management have replaced the two months of hospitalization experienced by the patients described in these early MGH records.

Some treatments in use in that time period are still in use in dermatology today. For example, silver nitrate, used for ulcers in Patients #15 and #72, is still in common use in the form of silver treatment and silver-impregnated dressings. A recent Cochrane review and meta-analysis found that silver-impregnated dressings improve reduction in wound size, but there is less data about long-term effects on wound healing.[8]

Mercury, which was proffered as a treatment for Patients #31, #72, and #78, has been in dermatologic news of late as a potential toxic additive to skin creams. According to the Environmental Protection Agency, several states have investigated cases of mercury poisoning due to the use of unlabeled skin lightening creams imported from Mexico and the Dominican Republic.[9] A recent U.S. Centers for Disease Control report warns that they have identified cases in both California and Virginia where patients with high levels of mercury were using creams containing thousands of times the level allowed by the U.S. Food and Drug Administration.[10] We wonder about the long-term effects that mercury pills, ingested by our patients in the 1820s for their skin complaints, might have had on their mortality.

Dr. Esther Freeman

---

[6] McCall Anderson T. *A Treatise on Diseases of the Skin*. London: Charles Griffin and Co, 1887.
[7] Hunter, J.A.A. "Turning points in Dermatology during the 20th century." *British Journal of Dermatology*, 2000; 143:30-40
[8] Carter, MJ, Tingley-Kelley K, Warriner RA. "Silver treatments and silver-impregnated dressings for the healing of leg wounds and ulcers: a systematic review and meta-analysis." JAAD, 2010 Oct;63(4):668-79.
[9] U.S. Environmental Protection Agency: http://www.epa.gov/hg/consumer.htm#creams Accessed 2/2/12.
[10] U.S. Centers for Disease Control and Prevention (CDC). *MMWR Morbidity and Mortality Weekly Report*, 2012 Jan 20; 61:33-6.

Patient #63: Scapula bruise

THEN

A German man entered the hospital on February 10, 1823 after a fall on the ice the day before. His pain was about the olcromian process of the scapula. He could not raise his hand to his head and only a short distance horizontally from his body. He was treated with hot packs. On February 12 he was doing well and on the 25th he could move his arm easily and was discharged "cured."

NOW

Minor injury

Patient #64: Childhood cataracts

THEN

A woman from Hanover, Massachusetts entered the hospital on February 10, 1823 with a female infant, five months old, who for two months had been afflicted with cataracts in both eyes. Her mother states that a short time before this, it was perceived that the child had an ophthalmic condition of which she was cured. Cataracts were now the color of the conjunctiva, which was a light bluish-white. The patient had strabismus and full contraction of the iris on exposure to light. The child's health, as well as that of the mother, was good. Belladonna was applied to the eyelids. She had full dilation of both pupils. The belladonna was then washed off with warm water. No perceptible diminution of the pupils took place, more than fifteen hours afterwards. On February 11 medication was applied with a tincture of opium without any appreciable effect. They did an operation on her eyes and belladonna was applied to the eyelids. Three drops of tincture of opium were given with a little water 20 minutes before the procedure. The head was firmly held by an assistant and the cataract needle passed into the globe a line distant from the external edge of the cornea. The cataract being a soft one it was punctured and the needle withdrawn. A bandage was next applied to the left eye and the head secured as before. Again the needle was placed in the opposite eye about the same place. The cataract was found to be firmer than the other. It was punctured and divided and the instruments removed. The child cried a little. Soon after, dressings were applied and the child became quiet. She was moved to a dark room. Post-operatively, she was given warm compresses. On February 15, she was doing well and on February 17 the eyes were not inflamed. She was discharged at her mother's request with a follow-up wash for the eyes.

NOW

This 5-month-old female patient presented with bilateral cataracts, which seem to have been first noted at the age of three months. It is unclear if the cataracts were congenital and not noticed until then, or if they were the result of a bilateral inflammatory or infectious condition (*"a short time, before this, it was perceived that the child had an ophthalmic condition of which she was cured."*) The cataracts may have become amblyogenic, since the patient was noted to have eye misalignment, as indicated by her inability to fixate properly. Lens opacities prevent clear images from forming on the patient's retinas and deprive the nervous system of the appropriate stimulation required for development of the visual pathways in the first months of life. The patient was treated with bilateral cataract surgery by prolapsing the cataract posteriorly in the right eye and division of the left cataract into pieces. This is essentially the same procedure as the one performed for Patient #51, a 75-year-old patient.

Pediatric cataracts still account for ten percent of visual loss in children worldwide, while about one in 250 newborns in the United States are found to have some type of cataract. Initial work-up today is aimed at determining whether the cataracts are congenital or acquired, isolated or associated with systemic or ocular conditions, and whether surgery to remove the cataract is indicated. A careful slit-lamp examination is performed in order to rule out associated abnormalities of the anterior segment of the eye, i.e., cornea, lens, iris, anterior chamber. If the cataract allows view of the posterior segment, a dilated fundoscopic examination of the retina and optic disc is also performed; otherwise, a B-scan ultrasound is obtained in order to rule out gross retinal and vitreal pathologies. Since unilateral cataracts are rarely associated with systemic diseases, a laboratory evaluation is not regularly pursued. In cases of bilateral cataracts without family history in otherwise healthy children, laboratory evaluation includes TORCH (toxoplasmosis, rubella, cytomegalovirus, herpes simplex) titers, and VDRL screening, analysis of urine for reducing substances and amino acids, and of blood for glucose, calcium, and phosphorus. Additional work-up is indicated if there are other developmental abnormalities and it should be done in consultation with a developmental pediatrician and geneticist.

Pediatric cataracts can involve the entire lens (lamellar or nuclear cataracts), such as in this patient, or only part of the lens structure (anterior polar cataracts or posterior lenticonus). The mere presence of lens opacity does not warrant surgery to remove the cataract; it is rather the visual significance of the cataract that will guide today's pediatric ophthalmologist in making a surgical recommendation (or not) to the patient's parents. Opacities that are central and greater than 3 mm in diameter are usually visually significant. Strabismus – such as in this patient - and/or nystagmus are signs that the optimal window for surgery and complete restoration of visual potential may have passed, though cataract extraction can still result in improvement of visual function. For bilateral visually significant cataracts, surgery prior to ten weeks of age is associated with a best-corrected visual acuity of 20/100 or better. Earlier surgery can result in even better visual acuity but the risk of aphakic glaucoma is increased.

Similar to adult cataract surgery, extraction (and not luxation into the posterior chamber as performed in the 1800s) is the preferred surgical procedure. Surgical technique is somewhat altered due to the elastic capsule of pediatric cataracts and ultrasound energy via phacoemulsification is generally not needed; the soft pediatric cataract can be removed using aspiration only. While implantation of an intraocular lens for children over one year of age has become the standard of care today, aphakia versus pseudophakia in younger children still remains a controversial issue due to the higher rate of re-operation and the variable refractive outcome.

Regardless of aphakia or pseudophakia after pediatric cataract surgery, post-operative follow-up is arranged for the first day after surgery at which time topical antibiotic, corticosteroid, and, sometimes, cycloplegic eye drops are started. Post-operative complications include primarily an increased inflammatory response of the pediatric eye to surgical manipulation with the potential formation of synechiae in pseudophakic eyes, the development of aphakic or pseudophakic glaucoma, and secondary opacification of the visual axis due to proliferation of remaining lens epithelial cells on the surface of the implanted intraocular lens. The rate of endophthalmitis and retinal detachment after cataract surgery in children is similar to adults.

Contact lenses are the preferred method to correct aphakia and prevent amblyopia in children after cataract extraction, while secondary intraocular lens implantation at an older age can also be performed in select cases.

While the hurdles facing nineteenth century eye surgeons were significant, they would pale in comparison to the challenges facing those who chose to do pediatric cases. The smaller anatomy and inability to communicate with young children mandates all such surgical procedures today be done under general anesthesia (first done in the Ether Dome at MGH in 1846). Subsequent post-operative exams would also have been difficult due to a lack of equipment and topical anesthetics which are now common.

Dr. James Chodosh
Dr. Matthew F. Gardiner
Dr. Joan W. Miller
Dr. Sotiria Palioura
Dr. Athanasios Papakostas

Patient #65: Iliac aneurysm

THEN

A male patient from West Newbury was admitted on February 12, having been in good health. About five weeks earlier he tried to lift some very heavy weight and although he felt nothing then, two weeks later he had a pulsating tumor developing in his groin. In a two-week duration, the tumor increased in size. The location of the tumor (aneurysm) was superiorly along the inferior margin of Poupart's ligament for 11 3/4 inches at its broadest part. Elevation of the tumor about the skin level, narrowing the tumor 1 3/4 inches measure at the highest level to about 1 1/4 below Poupart's ligament. It tapered off to an obtuse point, at hardly an inch in breath. When you put your hand on the aneurysm, there was a strong pulsation. When strong pressure was placed on the artery about the aneurysm, where it passes over the bone and extends into the thigh, the pulsation ceased entirely and aneurysm decreases to about half of its former elevation. Upon removing the pressure, the pulsation was renewed and the tumor grew to its original size in about six pulsations. The pain is aggravated in the horizontal position. At the time he was admitted he was in great pain. He was given a tincture of opium. On February 13 he had the hair from the tumor removed and a compress was kept on the tumor made of magnesium sulfate, tincture of menthol, and water. The patient complained of continued severe pain in the loins and he was given a tincture of opium and venesection, magnesium of sulfate, tincture of menthol and water. It was noted that on February 17, the aneurysm had increased in dimension in the previous 36 hours. He had severe pain in the back.

The patient required an operation because he could not live in his present condition. He wanted it done without delay. He was informed that many patients die during an operation but the patient was not dismayed. He was given a tincture of opium.

He was prepared for surgery. Scalpels, tenaculum, ligatures, several aneurysmal needles with ligatures, and blunt hooks were prepared. The patient was laid upon the table, head resting on a pillow. Two assistants secured his hands and arms and two others held the feet. Incision was made through the skin, about 3 1/2 inches long on a line drawn from the superior-anterior spinous process of the ileum to the symphysis pubis. Dissection was then carried out. Above the aneurysm was now laid bare and a small orifice cut through it. The fascia was divided in the direction of the wounds. Epigastric artery was found coming from the aneurysm sac. The sheath of the artery was exposed and a number of small veins from the thigh external to the artery were seen passing across the great artery to get to the vein. The veins were avoided. Incision was made into the sheaths of the artery to separate the path of aneurysmal needles, completely around and under the sheath. Ligature was firmly tied in a double knot. During the first part of the operation, a small, superficial artery was divided and secured without much hemorrhage. The patient endured the operation with considerable firmness. The wound was then sponged and brought together with adhesive straps, lint compresses, and the roller.

On February 19 the limb was warm and the back was less painful. He was examined and the aneurysm had diminished considerably. He was given magnesium sulfate and water and tincture of menthol. It appears he had numbness in his thigh above the knee, probably from the cutaneous nerves that were inadvertently divided. On February 20 he had had very little sleep. His limb was warm. Again, he was venesected. For nausea he received R-tartaric acid. On February 24 it was noted that he had "healthy" pus from around the ligatures. They used a bread and milk poultice over the wound. On February 28 it was noted that he had a slight discharge of "good pus" from around the ligatures. The ligature was removed on February 28 from the superficial artery. Ligature on the greater artery was firm. The aneurysm was diminishing gradually. He was discharging "healthy" pus in moderate quantity. The pus was coming from a small opening in the wound about an inch from the ligature and towards the external end of the wound. They applied a poultice of bread and milk.

On March 6, the separation had almost ceased. On March 14, he developed fever and chills with hot and cold flashes. On March 18 his discharge ceased. There was some "fungus" noted on the lips of the wound. Silver nitrate was applied. On March 24 he had oozing of blood from the wound. The blood was thought to come from the fungus on the lips of the wound. They applied hydrangea to the wound. On April 15 his wound was noted to be almost completely healed. He was finally discharged on April 28, "cured."

NOW

The patient described above appears to present with a degenerative aneurysm of the superficial femoral, or more likely, the deep (profunda) femoral artery. The temporal relationship between the diagnosis/presentation and the episode of heavy lifting may not be related. If they are, in fact, related, it is most likely that strenuous activity led to a dissection in the femoral artery, and this, in turn, led to formation of a pseudoaneurysm or a degenerative aneurysm of dissection etiology. On his sixth day in the hospital, February 17, it seems that his pseudoaneurysm ruptured into the femoral sheath, which tracked up to the retroperitoneum, causing acute worsening of his severe back pain.

At this point, he was appropriately taken to the operating room for aneurysm exclusion. The operative plan at that time was in keeping with the current standards. The aneurysm was dissected free of surrounding tissues, ligated proximally, and then the aneurysm sac itself was packed with steel wire, in an effort to cause aneurysm thrombosis. In the post-operative period, collateral circulation and the non-aneurysmal branch of the femoral artery were responsible for keeping the leg perfused. Some of the post-operative inflammatory response was likely due to thrombosis of the aneurysm, though it was also caused in large part by what was likely a wound infection ("healthy" pus draining from the wound). The documentation of the decreasing size of the aneurysm is consistent with aneurysm thrombosis and sac collapse.

In 2011, a similar patient would likely present with a pulsatile mass in the groin (with or without pain), and be initially evaluated with Duplex ultrasound and/or computed tomography angiography (CTA). These studies would reveal the anatomic boundaries of the aneurysm, and also delineate whether the diagnosis was a true of false (pseudo-) aneurysm. If the aneurysm sac were greater than 3 cm in size, or certainly if there were evidence of rupture (as in the case above), plans would be made for operative exploration and aneurysm resection. In the current paradigm, however, the goal of therapy would be aneurysm resection and return of blood flow, rather than aneurysm thrombosis.

Intravenous antibiotics covering usual skin flora would be given within an hour preceding the surgical incision. Rather than the oblique incision described above, the incision of choice would be longitudinal, allowing for distal exposure, as the aneurysm anatomy dictates. Electro cautery would be used to dissect down to the level of the inguinal ligament and the femoral sheath. Sharp dissection would then be used to dissect free normal segments of the common femoral, superficial femoral, and deep femoral arteries. All would be encircled with elastic vessel loops. As described in the case above, in the dissection of distal portion of the deep formal artery, one encounters multiple crossing branches of the femoral vein, and one must take care to avoid injury to the branches of the femoral nerve.

To prepare for exclusion of the deep femoral artery aneurysm, after the administration of systemic anticoagulation, vascular clamps would be applied to the common femoral, superficial femoral, and deep femoral arteries in segments where they are morphologically normal. Now, in a bloodless field, the aneurysm would be opened longitudinally with a knife, and this opening would be extended with Potts' scissors cranially and caudally, into normal-appearing artery. If possible, the artery would be completely transected at the proximal and distal extents of the aneurysm, and the artery would be replaced with an 8 mm woven Dacron (or ePTFE) graft. A simple in-line reconstruction would be carried out if the proximal extent of the aneurysm did not involve the femoral bifurcation, sparing the origin of the superficial femoral artery. If the femoral bifurcation were involved, then the reconstruction would require use of a bifurcated graft, or a re-implantation of the normal femoral artery onto the Dacron graft. After completion of the arterial replacement, the clamps would be removed, and the adequacy of the repair would be assessed with palpation, Doppler interrogation, and pulse volume recordings of the lower leg. The choice to close the femoral sheath over the repair depends on the surgeon. The layer of Scarpa's fascia and the subcutaneous layers of tissue would be closed with absorbable sutures, and the skin would be closed according to surgeon preference.

Post-operative management would consist of aspirin therapy, and graft surveillance would be completed with daily pulse volume recordings. Peri-operative antibiotics would be continued for 24 hours. Should a wound infection be noted, systemic antibiotics would be instituted for a seven-day course. After 24 hours of bed rest, the patient would be allowed to ambulate. Discharge from the hospital would then be expected on the second post-operative day. The patient would then return as an outpatient in four to six weeks for wound evaluation and for graft evaluation with duplex ultrasound.

Dr. R. Todd Lancaster
Dr. Virendra I. Patel

Patient #66: Trigeminal neuralgia

THEN

A 71-year-old male came to the hospital February 18, 1823. He complained that fourteen years earlier he acute pain which was confined to the lower jaw. Paroxysm lasted about a minute, and the pain was so severe that he compared it to the boring of an auger while tearing the flesh forcibly from the bone.

The intervals of the contractions were ten days to two weeks. They became shorter and shorter. The pain for the past three months had been constant, but less severe than before and it was only on the right side. He had the inferior maxillary nerve divided with some relief. He had taken large doses of stramonium and other narcotics.

An operation to divide the facial nerve at its exit from the cranium to the foramen silo mastoid was performed as follows. An incision was made about three inches long beginning in the front of the mastoid process of the temporal bone as near as possible to back part of the ear, extending down the neck in a course of the inferior edge of the sterno mastoid muscle. The incision was carefully continued under the base of the cranial nerve which was discovered with some difficulty owing to the blood which flowed from the division of the occipital branch of the external carotid. The nerve was then divided but without any increase of pain to the patient rendering it somewhat doubtful. Knife was then frequently carried over the foramen in order to remove every possibility of the nerve's division. The wound was dressed with adhesive straps.

On February 19, his pain continued the same as before the operation. He had a paroxysm of pain that morning. When he had the pain in the lower jaw, his speech became difficult and sometimes suspended until the paroxysm subsided. He refused to take any other narcotic or opium. On February 25, it was noted that the pain had continued with the same severity. On February 26 an operation was performed. For the operation, the patient took three grains of opium. An incision was made just below the zygomatic process of the temporal bone and over the body of the inferior maxillary bone on the right side through the skin near the parotid gland and the masticator muscle. The dissection was carried through the gland, dividing the duct and then carried to the muscle, through the bone without dividing any considerable artery except one which was secured by ligature. The great facial artery was exposed in the upper part of the wound. The periosteum was removed. A dressing was then applied and the bone perforated just opposite to the foramen as was shown by furrow niche in that portion of the bone which was removed bringing with it a portion of the nerve with one of its branches. The remaining part of the nerve was then drawn out by the forceps and cut off at the foramen. The other end had a piece removed a little more than a half inch long. The edges were drawn together by a crooked needle ligature and securing them by a knot. The wounds were then dressed with adhesive straps and the patient was walked to his room. He was pretty well satisfied that the desired relief was obtained.

There was some bleeding from the superficial artery which was treated by compression. He thought he could do without the opiates at night. Probing the wound at the level of the bone did not hurt. However, when it was moved about one half inch from the bone he had severe and very acute painful paroxysms which almost suspended respirations but subsided in less than a minute and a half.

On March 5, he had some discharge of pus from the wound and probably some saliva also. The jaw bone seemed sore. On March 8 he had difficulty swallowing and some redness around the right tonsil and roof of the mouth. On March 10, his throat was quite sore and painful. He was given muriatic acid and water, and simple syrup. An abscess burst in the right tonsil on the night of March 13. On March 15, he noticed that his old pains had gone. He was discharged, considered "cured."

NOW

The patient appears to have had right trigeminal neuralgia (tic douloureux) involving the third division, possibly associated with hemifacial spasm (tic convulsif), as the two conditions can occur simultaneously.

Trigeminal neuralgia has been recognized for centuries and is characterized by intense, lancinating, sharp electric-shock-like pain in the trigeminal nerve distribution, predominantly second and third divisions. As these nerves also supply sensation to the teeth and jaws this condition is often initially confused with dental disease or jaw pathology. The pain can be triggered by movement of the face or jaw, touching the face, brushing the teeth or talking. It is sometimes associated with the patient wincing in pain and a distortion of the face. Between attacks the patient is pain free with no other neurological signs. The pain goes into remission for a variable length of time but usually returns.

Hemifacial spasm is an involuntary twitching or contortion of the facial muscles on one side of the face and is not always painful. It can be caused by compression of the VII nerve as it exits the brainstem by the anterior inferior cerebellar artery or a tumor or cyst in the same area. In the case of trigeminal neuralgia, the superior cerebellar artery can compress the V nerve. An arachnoid cyst or meningioma can also compress the nerve at the cerebellopontine angle.

The patient had three procedures. The first was to divide the "inferior maxillary nerve," possibly the mental nerve which provided some relief. The second procedure, apparently performed without analgesia, divided the facial nerve. The third procedure, accomplished with three grains of opium for analgesia, divided the inferior alveolar nerve leading to post-operative infection and spontaneous abscess drainage. After this the patient was considered to be cured and was discharged. He would have had right facial paralysis and numbness of the right lower lip.

Imaging with readily available dental films, such as the panoramic radiograph, would quickly rule in or out the presence of an infected impacted tooth, odontogenic lesions or tumors, or inflammatory jaw conditions.

Patients with trigeminal neuralgia (TN) or hemifacial spasm undergo imaging studies: MRI or MRA to assess for intra-cranial pathology or blood vessel impingement. Treatment of TN consists primarily of anti-seizure medications. Neurosurgical interventions include microvascular decompression, radiofrequency rhizotomy, balloon decompression or glycerol injection into Meckel's cave. Gamma Knife radiosurgery has also been used. Peripheral neurectomies are rarely recommended as TN is a central nervous system condition and the pain frequently returns in the numb area causing anesthesia dolorosa. Hemifacial spasm is likewise treated with anticonvulsants, muscle relaxants, botulinum toxin injections, or microvascular decompression. Odontogenic infections would be treated with antibiotics and removal of the offending tooth. Jaw cysts or tumors or other pathology would be excised.

Dr. Leonard B. Kaban
Dr. David A. Keith
Dr. Thomas B. Dodson

Patient #67: Frostbite of toes

THEN

A 35-year-old male entered the hospital February 21, 1823. Four years earlier he had had the toes of both feet frozen. Shortly after the toes were removed and healed, ulcers had appeared on his legs which were now open and much inflamed. His leg ulcers were treated with poultice of rye meal and milk with carrots three times a day. On February 28, the ulcers appeared to be healing. His gums were a little sore. The patient left during treatment. Condition: eloped.

NOW (See Patient #16)

Patient #68: Tibial fracture

THEN

A 43-year-old male was admitted on March 6, 1823, with an oblique fracture of the left tibia and fibula about five inches from the ankle joint. He fell stepping through ice and then when pulling his leg out he fell again, breaking it. The obliquities go from above downwards from within outwards. A small piece of bone lay loose upon the spine of his tibia on the upper part of the fracture. The limb was swollen; doctors tried to reduce the fracture. The extremities of the bone could not be easily retained in a natural situation. Eighteen tail bandages were applied, moderately tight, the limb being supported by a pillow and splints. The tail bandages were secured by tapes and a cradle placed over the limb. On April 2, it was noted that the limb was not perfectly united and there was some motion between the extremities of bone. On April 15, the limb was very straight but there was still some play between the broken portions of the bone. The extremities of bone were pretty thoroughly united. On April 29, it was noted that the bones were perfectly united, but the knee was pretty stiff. On May 3, the limb felt strong, the knee was more limber. On May 20, the leg continued to swell at night upon the least amount of exercise. On May 26 he was discharged.

NOW (See Patient #7)

Patient #69: Cataract

THEN

A 36-year-old male was admitted April 2, 1823, with a cataract of the right eye. In September his cornea was scarred and at that time he had no pain but his vision was blurred. The cataract had a blue pearly luster. Operation consisted of pre-operative application of belladonna. A needle was passed into the globe two lines distance to the edge of the cornea. The cataract was broken with the needle. On April 23 the cataract appeared to be dissolving and the membrane was less opaque.

NOW

This case describes a 36-year-old male with cataract and corneal scarring in his right eye. The blue color of the cataract may represent a description of a Cerulean cataract which is characterized by the presence of tiny dots or flake-like opacities which are typically organized in a radial configuration. The Cerulean cataract, which was first described by Vogt in 1922, appears to have predominantly bluish-green or white opacifications when viewed by slit-lamp biomicroscopy using direct illumination. The corneal scarring is likely secondary to old trauma which might have led to the formation of the cataract. The scarring could also be attributed to a variety of other diseases such as trachoma (leading cause for corneal blindness in the nineteenth century), prior episodes of herpetic keratitis, or even apical scarring in keratoconus.

Modern treatment of cataract surgery is phacoemulsification surgery which involves the use of an ultrasonic handpiece equipped with a titanium or steel tip which vibrates causing the lens material to be emulsified. A second fine instrument (sometimes called a "cracker" or "chopper") is also used through a second incision to facilitate breaking the nucleus into smaller pieces for aspiration out of the eye. After phacoemulsification of the lens nucleus and residual cortex is complete, an artificial lens made of acrylic or silicone is implanted in the eye to maximize the vision without spectacles or contact lenses. Anesthesia for cataract surgery is now done by local injection and sedation or with topical agents such that patients are completely comfortable. See Patient #51 for further discussion of cataract surgery.

Dr. James Chodosh
Dr. Matthew F. Gardiner
Dr. Joan W. Miller
Dr. Sotiria Palioura
Dr. Athanasios Papakostas

Patient #70: Knee bruise

THEN

A 23-year-old female was admitted on April 10, 1823. She had slipped on the ice and fallen on her knee. It was sore for a few days and not very painful until she heard a snapping of the joint and was now lame and could not use her joint. A blister was applied. On April 24 she had weakness of the joint, but by April 29 she was feeling well and was discharged.

NOW

Minor orthopedic condition

Patient #71: Chronic loss of sight

THEN

A 26-year-old blacksmith was admitted on April 13, 1823. He had stopped working eight years earlier due to a gradual loss of sight. He had pain with light and was very sore. He had blisters applied behind his ears. On July 11 he was discharged.

NOW

Although gradual painless loss of vision can be the result of various disease processes that affect either the anterior (chronic corneal disease or cataract) or posterior (chronic retinal or optic nerve disease) segments of the eye, there are only a handful of conditions that cause painful visual loss. The differential diagnosis of *acute* painful visual loss includes acute angle-closure glaucoma, endophthalmitis, corneal hydrops or ulcer formation, and optic neuritis. Regarding *chronic* painful visual loss, recurrent optic neuritis and uveitis (inflammation in the anterior and/or posterior segment of the eye) are the most common causes. Though this 26-year-old patient belongs to the age group that optic neuritis usually affects, he is also experiencing severe sensitivity to light ("*Had pain with light and was very sore*"), which is a symptom commonly caused by intraocular inflammation. Thus, this patient is most likely suffering from chronic uveitis.

The modern work-up for any patient with uveitis is aimed at identifying if the source of inflammation is infectious, rheumatologic, or part of a masquerade syndrome (e.g., amyloidosis, ischemia, chronic retinal detachment, occult foreign body or neoplasia). Apart from a complete ophthalmic examination, the initial evaluation should be tailored based on the course and duration of the patient's symptoms. Infectious uveitis that requires specific laboratory testing includes syphilis (anti-treponemal serology and/or cerebrospinal fluid serology), tuberculosis (tuberculin purified protein derivative skin test), Lyme disease (Lyme serologies), and toxoplasmosis (toxoplasma serologies). On the other hand, if the cause of uveitis is a granulomatous systemic disease (e.g., sarcoidosis, Wegener's granulomatosis) the diagnosis may not be established until after several years when systemic symptoms from the disease make it obvious. If there is no evidence of infection, neoplasia, or other masquerade syndromes, uveitis is thought to be the result of an autoimmune process (e.g., Vogt-Koyanagi-Harada disease, birdshot chorioretinopathy, sympathetic ophthalmia, chronic anterior uveitis due to juvenile idiopathic arthritis or HLA-B27 associated disorders, Behçet disease, pars planitis).

If an infectious or neoplastic etiology is found, appropriate antibiotic or chemotherapeutic therapy is started. If, however, the cause is presumed to be autoimmune in nature, immunosuppressive (i.e., topical and/or systemic corticosteroids) and immunomodulatory agents are used. Immunomodulatory agents include the anti-metabolites (e.g., methotrexate, mycophenolate mofetil, azathioprine), T-cell active agents (e.g., cyclosporine, tacrolimus), cytotoxic alkylating agents (e.g., cyclophosphamide, chlorambucil), and biologic agents (e.g., TNFα-inhibitors). Vision-threatening complications from chronic uveitis include cystoid macular edema, vitreous opacities unresponsive to medical therapy, posterior synechiae, increased intraocular pressure, and very low intraocular pressure (hypotony).

Uveitis is a chronic disease and, thus, the duration of therapy is usually long. Moreover, due to the many sight-threatening complications, most uveitis specialists continue therapy for at least one to three years after disease quiescence. Follow-up is individualized and can be from every few days if the inflammation is severe and high-dose corticosteroids are used to every three to four months if the disease is stable and well-controlled. (See also discussion of Patient #57).

A common treatment at this time was the creation of "blisters" on the skin by the use of irritating chemicals, such as Blistex, in an attempt to draw out the "evil humors." This may explain the reference to blisters behind the ears in this patient and is also seen in Patients #51, 57, 82, and 92. Similar concepts justified the use of leeches or cupping (applying cups filled with warm air to the skin to create suction).

Dr. James Chodosh
Dr. Matthew F. Gardiner
Dr. Joan W. Miller
Dr. Sotiria Palioura
Dr. Athanasios Papakostas

Patient #72: Syphilis

THEN

The patient was a 24-year-old male hairdresser from Boston. He had contracted syphilis nine months earlier, and soon after had two chancres on the extremity of his penis. (Previous to that he had had gonorrhea.) These disappeared in about a month. The chancres were accompanied by two swellings in the groin, one on the right and the other on the left. The right swelling was removed, the latter remained and was lanced. Ever since it was lanced, it continued to discharge copiously and now was considerably swollen, inflamed, and very painful. There was swelling between the upper part of the thigh and the scrotum which he complained about, saying that the pain was very severe and it interfered with his walking. This increased to great size and finally broke open, and the discharge was irregular and foul. He had several indurated lumps under the inner part of the thigh and two ulcers just above, but were smaller, superficial, and in a healing state. The abdomen was very swollen.

He complained of pain in his bowels which were rather loose. His general health was pretty good although he was somewhat emaciated and suffered great pain. He had been under a course of treatment of mercury and had a sore mouth. Various topical applications were made use of, none of which aggravated the disease. He had warm baths and washed all the parts with soap and water. Linen rags kept the ulcers clear until morning. He got hydrangea and opium ten grains. On April 15, he was visited by a doctor who prescribed hydrangea and pulverized opium. He took this at night and in the morning. He washed daily two or three times and dressed with simple ointment. He had a milk diet, with no animal food. The patient thought that the milk did not agree with him. They decided to give him three opium pills per day. The wounds were dressed with simple ointment. The ulcers started to look better, less inflamed. On April 29 the abdomen was less swollen, and bowels regular. He omitted the wash and applied poultices and boiled carrots three times per day. By May 3, he started to improve. Medication was omitted as were the mercury pills. He continued the poultices for a few more days, three times per day. On May 9, he continued to apply silver nitrate. His wounds were treated with a caustic agent. The mass on the inside of his thigh was almost well. On May 19, his wounds were healed and he was discharged "cured."

NOW (See Patient #27)

Patient #73: Migratory arthritis

THEN

A 27-year-old male from Newburyport entered the hospital on April 19, 1823. He was a silversmith. He had been married four years and had three children. Four years earlier he had rheumatic pains from his shoulders through to the breast, accompanied by difficulty breathing when walking. He had never had any sickness before until six years earlier he had fevers and dysentery and was confined about three weeks. He now had developed rheumatic pains in hip. Motion of the hip increases the pain. He used to walk a half mile to his shop twice a day. He bathed these sore parts with oil and turpentine without any benefit. He had severe sweats at night but slept mostly undisturbed except when in great pain. He had great pain when coughing or sneezing, as though he was pulling the hip out of his joint. It was recommended that he have a sulfa bath every night. Note: his pain was originally in his shoulders and through the breast. He also had, for the last four years, pain in the hips. When he was discharged his pain was less acute.

NOW

Non-surgical condition

Patient #74: Musculoskeletal infection

THEN

A male patient, age 11, was admitted with swelling of the knee and leg. The first appearance of this was about two years earlier, and since that time had been growing. He had been using the leg with no great deal of inconvenience until the swelling ruptured a year before his admission, just above the internal condoyle. There continued to be discharge at time of admission. He had been under the care of physicians and made use of remedies and adhesive plaster. The motion of the knee was limited and could not be bent to form a right angle. It was much more swollen than usual. The swelling extended from above the knee down to the ankle. The calf on the leg is considerably enlarged, increased three inches more than the well one; above the thigh two and a half inches more than the other. He experienced pain and soreness on the under part of the thigh, and it was very tender and painful upon touching. He said he had never had any fragments of bone discharge. He has fine skin and a light complexion. Bowels were open and appetite was tolerable. The area of the leg was dressed with ointment twice a day. On April 23 he had further examination and it was found that the thigh had been affected and now it also had a little swelling and tenderness. He said it had been considerably more sore, painful, and swollen. The bones were also diseased. On May 3 the swelling below the knee subsided to some degree and increased above the knee. It was very painful. It was now black. On May 5, the swelling was decreased as was the pain. A blister was opened, the discharge continued, his pulse was slow and feeble. Blisters kept being applied above the knee.

Blisters were open, discharge from sores was considerable, and there was discharge from blisters. He walked around without great difficulty. By July 18, he had scarcely any pain in his limb. He had diminished doses of medications. Blister was nearly gone. On July 25, he was much the same. On August 6, the inner mass began to discharge. He generally felt better and stronger. On August 8, he was about the same in all respects. He was discharged and general health was "pretty good."

NOW (See Patient #35)

Patient #75: Osteomyelitis

THEN

A 19-year-old male came to the hospital April 24, 1823 with lower leg pain. He remembered that on March 5, 1819, he suddenly had had a pain in his leg. The day before he had gone to a meeting and it was very cold and disagreeable weather. The pain gradually increased until he developed a fever and was deprived of his senses. He compared the pain in his limb to violent cramps of rheumatism. At times it was so severe as to affect his respiration. The pain first came in the middle of the tibia and was deep, and there was no appearance of inflammation on the skin at that time. His confinement with fever lasted about a year. His leg was greatly inflamed and enlarged until it broke open and discharged freely, giving much relief. This was in the early stages of febrile action. Ever since it had remained open in some places, frequently heals then inflames, swells, and breaks apart in other places.

The surface of the bone has been exposed and exfoliated and small pieces of dead bone have been discharged from time to time. In the last year his leg had not been very painful nor had it changed in appearance. It drained from a fistula on his tibia. The center of the tibia was found to be rotten when probed and dead in different places. He used medications to bathe the area. It did not heal. An operation was performed. The equipment for the operation included: probes of different sizes, mallet, gauge, chisels, trephine, A-saw, bone nippers, cold water ligatures, lint pledgets, and compression bandages.

The patient was held down by assistants. The surgeon examined with a probe, looking for the extent of dead bone and for the firmness of the living bone. He commenced the procedure by making a long incision through the skin of about four inches, which he crossed with another two inches, and then carefully dissected the skin, exposing the bone, which proved to be considerably thick and firm. The necrosis was deep in the bone, covered over with bone tissue, and just opened at the fistula site. The instrument was applied to the external surface of the bone. The trephine was used and the chisel gauge was elevated to the junction of the living bone, and a portion of the living bone was taken out. Part of the dead bone was laid bare. The probe was again passed for further examination. The cavity was ascertained to be filled with old bone. The operator proceeded to remove the newly formed bone making use of the saws, gauge, chisel, and whatever the particular circumstance required. At the same time, the incision through the skin was carried out further, made crossways on the flaps, and then dissected up. A good part of the sequestrum was now visible and it was difficult to remove it without breaking it. This was done with A-saws, and two portions of the bone were extracted without any difficulty. The consequences of the ragged edges of the dead bone being strongly intermingled with new formed bone, excessive portions of normal bone were again and again removed by the same instruments until the whole sequestrum was extracted and the operation completed. The dead portion of bone was found to extend nearly the whole length of the tibia and a new formed bone sufficiently gave strength to the limb. The bone was then cleaned and filled with lint, over which were applied pledgets. The patient was brought to his bed completely exhausted. During the operation his strength was supported with wine. The operation took about two hours and was borne by the patient with fortitude. Hemorrhage was not much as to attract any particular notice. The pieces of bone removed were six and seven inches in length. He was given opium intra-operatively and post-operatively. He was also given a mixture which included some sulfuric acid, as well as mint tea. On May 9, the third post-operative day, some sloughing was seen at the edge of the wound. On May 11, the drainage in the leg was very disagreeable. The wound was cleaned and the lint was removed. On May 12, there was a slight discharge from the leg. On May 13, the wound appeared healthy in appearance with the cavity covered with healthy granulations, and separation going on very well. The wound was then dressed twice a day and he had been taking charcoal but it was omitted. On May 15, nine days after the operation, he had normal appetite and increased feeling in the leg. The cavity was still filled with pus and the surface looked good. Some portions of bone were exposed.

On May 18 he was able to sit up and go about the room. There was copious suppuration. By May 27, suppurations continued freely. Another orifice opened up to discharge and it was irrigated with magnesium sulfate. On June 10 he was considered to be doing well, feeling stronger, and the suppuration was less. The edges were contracting and bone was recovering very slowly, as noted on July 4. Also it was noted continued increase in exfoliation of two pieces of bone which appeared to be part of the tibia.

On July 16 the operation was repeated in the upper part of the tibia, near to the head. At first an attempt was made to remove this portion of bone with forceps but it was ineffectual. Diseased bone was so completely and thoroughly enveloped that it was impossible to take it out. The session proceeded to the use of the scalpel, trephine, saws, chisels, with the same general principles of the former operation. The cavity of the dead bone was about three inches deep. The pieces of dead bone were taken out and the cavity was cleaned and filled with lint and covered with compresses. The patient bore the operation very well. He was given a tincture of opium. The discharged increased.

The July 16 operation was done to remove the dead bone by removing living bone about two inches by two inches. Charcoal was applied. On September 2, the wound was filling up. Naked portions of bone were not exfoliated. They were suppurating freely between September 8 and September 16. He was doing well. The wound was rapidly filling up with flesh and strength. On September 19 he was doing well with no evidence of any diseased bone. Every part seemed perfectly healthy. They thought it was going to be a perfect cure. The patient was discharged on September 20, 1823.

NOW (See Patient #35)

Patient #76: Child with residual cataract

THEN

The patient, an infant child who had had a cataract in the right eye which was wholly cured, had been previously been admitted on February 10, 1823. She was then readmitted on April 28, 1823. The pupil was clear and sound. The cataract to the left eye was partially dissolved and an opaque spot still remained. Belladonna was applied. On April 29, the day after her admission, the child was held by a female sitting in a chair, the retractor was put in with a speculum and the surgeon passed into the eye and punctured the cataract. During the operation, the child had a lot of resistance and cried considerably.

Her upper eyelid was inverted, which together with the tears from the eye almost entirely obstructed the site of the operation. Occasionally, however, the opaque spot appeared to view and was much lacerated. Small compresses of lint were applied with a bandage wound around the head. In a short time, the child was perfectly well. That same night the child appeared cheerful and was discharged on April 30. She was in the hospital for two days, one day pre-operatively and one post-operatively.

NOW

This patient was readmitted since pieces of the initial cataract were still seen in the anterior chamber of the left eye about two months after the initial operation. Unlike Patient #51, who experienced an inflammatory response and a rise in intraocular pressure due to the retained lens fragments in the anterior chamber, there is no evidence in the clinical history that this patient developed such signs and symptoms. It seems that the second operation in this case was prompted by the mere presence of pieces in the anterior chamber two months post-operatively.

Currently, instead of being displaced into the posterior chamber the cataract is extracted usually by phacoemulsification (see Patient #51). Thus, retained lens fragments in the anterior chamber are not commonly seen post-operatively. Today, in the absence of inflammation or rise in intraocular pressure, such lens fragments are followed up until dissolution, which may take several months. Surgical intervention to remove lens fragments is clearly indicated if they result in anterior chamber inflammation, a rise in intraocular pressure or corneal edema. Follow-up every one to two weeks for measurement of the intraocular pressure until any retained lens fragments dissolve is generally indicated.

In the nineteenth century, physicians were unable to completely examine the eye in order to search for residual lens material in the posterior segment. In the event that posteriorly located pieces of the cataract did not dissolve, chronic inflammation would result. Unfortunately, steroids for the control of inflammation were not available at this time. Nevertheless, knowledge of the presence of retained lens material would have been of little benefit since techniques for vitreoretinal surgery were over 130 years in the future (developed by Dr. Charles L. Schepens in the 1950s).

Dr. James Chodosh
Dr. Matthew F. Gardiner
Dr. Joan W. Miller
Dr. Sotiria Palioura
Dr. Athanasios Papakostas

Patient #77: Hemorrhoids

THEN

A 64-year-old male was admitted on April 28, 1823. He had a 20-year history of "piles" and of constipation. A physical exam revealed two small tumors, one on each side of the anus. There was no inflammation. He was treated with magnesium sulfate and bougie in the rectum every night, increasing the size daily. From May 5 to May 11, he received laxatives and bougies and was released on May 11, much relieved.

NOW

This patient had hemorrhoids, which were treated with anal dilatation and laxatives. Today, most hemorrhoid patients are treated simply with fiber and warm water soaks. If intervention beyond this is desired, many patients can be treated in an office setting with rubber band ligation or infrared coagulation of internal hemorrhoids. Large and/or external hemorrhoids are treated with formal hemorrhoidectomy under anesthesia.

Dr. Paul C. Shellito

Patient #78: Syphilis

THEN

A male, age 24, was admitted on May 2, 1823 for his second bout of syphilis, his first being six years earlier. Upon being admitted, he had had it for eight weeks. He was taking mercury pills and his mouth was sore. Initially the patient had gonorrhea and two chancres on the glans. The penis was enlarged and inflamed. On May 3 it was noted that he had poor appetite. On May 13 he received a mercury pill, and on May 20 the chancres were looking better. On May 26, a caustic was placed under the prepuce daily and zinc sulfate was used. He was allowed two glasses of wine per day and a house diet. He had slow improvement and was discharged on June 19.

NOW (See Patient #15)

Patient #79: Clinical case – unreadable

THEN

A 33-year-old Spaniard was admitted on May 23, 1823. The rest of the patient's case history is illegible.

NOW

No commentary

Patient #80: Femoral fracture

THEN

A 48-year-old male pump maker was admitted on May 12, 1823. He fell while intoxicated. He was violently pushed through a door and down three steps, and hit the pavement, fracturing his upper thigh. He was unable to help himself in any way. Although in excruciating pain, he did not go to the doctor until the next day. The fracture was set, but no fracture box was applied. At 8:00 p.m. on the day of admission, one leg was three inches shorter than the other. The toes and knee of the affected leg were turned outwards and there was swelling around the groin and syphilis pubis. Within two inches there was a considerable prominence which appeared to be protruded by the head of the fractured portion of the bone. Extension and counter extension were gradually applied for a considerable time for the limb to come into place. The muscles were violent and spasmodic; a large fracture box was applied. The patient was delirious. Three hundred drops of opium were given. On May 28, his pain decreased and by May 31 was delirious only on and off. On June 20, he deteriorated and died.

NOW (See Patient #8)

Patient #81: Minor foot trauma

THEN

Patient is a 21-year-old male who came to the hospital on May 24, 1823. He had recently received an injury to the instep which produced slight spasms, and then he was run over by the wheel of a wagon which resulted in severe spasms of this foot and leg, and shrinking of his leg leading to atrophy. He cut off two of his toes with a chisel, thinking that that was the source of his pain. He said the foot felt cold at times. The right foot is smaller and flabbier than the left. The right leg is about a half inch shorter than the left and measures at the instep a quarter inch less at the ankle, one inch less above the calf. He was advised to abstain from spirits. He had fits of epilepsy which continued to affect him. The attacks of epilepsy last from one to four days at intervals of one to three months. At the time of patient's admission, he stated that his current attack had commenced two days earlier, and that prior to that attack, he had not had one for fourteen weeks. On June 5, he felt as though he was having a return of his paroxysm. He was having agonizing pain at the base of his knee. He had paroxysms and did not take any nourishment. He felt that the medications did him good. He went on poultices and a drug. He was discharged for bad conduct, apparently "cured."

NOW

No commentary

Patient #82: Severe conjunctivitis

THEN

A 17-year-old woman was admitted to the hospital on May 24, 1823 with a scrofulous inflammation of her eyes. She had a five-week history of pain and inflammation of the eyes, which got worse in the afternoon. She had significant secretion from the eyes. She woke up in the mornings with her eyes glued shut. It was often difficult to separate her lids. She had been on medical treatment and made use of many blisters, applied daily, with small portions of salts. She had an ounce or two of blood taken from her arms until the physicians were unable to withdraw any more. For about a year, she had taken magnesium sulfate. On the second day of her hospital stay, May 25, she did not sleep. Leeches were applied to her temples and opiates given. No animal food given.

On May 26, the inflammation to her eyes was greatly improved. A small compress with some plumbi liquid and aqueous solution compresses were constantly applied to the eyes. On May 27, the pain was very bad, though the inflammation was less than the day before. She did not sleep much. On May 29, she had an uncomfortable day, and vomited after ipecac. Her eyes stayed inflamed. On May 30 her eyes were no better. The pain was very severe. Medication was placed on her neck. On May 31, the pain was as before. She used aspergilla and tincture of oximus. She was given an application of burnt alum to the eyes. Her eyes generally improved.

She was noted to have scrofulous tumors along the neck for a few days. Eight leeches to the temples were applied with no great relief from either the blisters or leeches. She had venesection without relief. Twelve leeches were applied the next day. On July 4, she wished to be discharged, but had not improved.

NOW

This is a description of a patient with severe conjunctivitis, which can be caused by a variety of agents. Bacteria and viruses are the most common etiologies with streptococcus and adenovirus being predominant. Another possibility for this patient is phlyctenular keratoconjuctivitis (PKC), most probably secondary to *Mycobacterium tuberculosis*, given the presence of "scrofules" (*scrofa* is the Latin word for brood sow, which children with lymphatic swellings in the neck were thought to resemble). PKC is a localized noninfectious inflammatory/ hypersensitivity disorder of the ocular surface characterized by subepithelial nodules of the conjunctiva and/or cornea. The pathogenesis of PKC is thought to be a type IV hypersensitivity reaction to an antigen of bacterial origin. PKC has been classically associated with *M. tuberculosis* (especially in developing countries). However, *Staphylococcus aureus* is the cause in a majority of cases in the United States today. Given the possibility of TB, a PPD test and chest x-ray would be warranted as well as a complete physical exam.

Topical antibiotics are given empirically for suspected cases of bacterial conjunctivitis, while viral cases are treated with symptomatic measures and supportive care, and are self-limited. Culture is often very helpful for refractive cases or for patients at risk for or with signs specific for chlamydia and/or gonorrhea. Topical corticosteroids are the mainstay of treatment for phlyctenulosis. The lesions are exquisitely sensitive to topical steroids, and one drop of topical one percent prednisolone twice daily is generally effective within 48 to 96 hours. Concomitant topical antibiotic therapy and lid hygiene to control any associated blepharitis is mandatory. The application of warm compresses is likely the only therapy delivered to this patient which provided any benefit during her hospital stay. Over the course of a one month admission she received bleeding, blisters, leeches, ipecac, and burnt alum applied to the eyes. Having found no relief, she requested discharge (perhaps against medical advice).

Dr. James Chodosh
Dr. Matthew F. Gardiner
Dr. Joan W. Miller
Dr. Sotiria Palioura
Dr. Athanasios Papakostas

Patient #83: Partially dissolved cataract

THEN

Patient #69 was readmitted on May 25, 1823 with a partially dissolved cataract after the previous operation. The right portion appears clear. A few pieces of cataract were floating in the eye. Near the inside of the pupil a portion of cataract is in an unfavorable place for dissolving. An operation was advised. Belladonna was used to dilate the pupil. The operator passed a needle to the globe of the eye two lines from the cornea. The pupil was cleared. After completion the needle was withdrawn and a compress placed. On May 29 he continued to complain of cloudy vision. A small portion of the lens capsule on the right was found to be unabsorbed and was connected to a small portion of the iris at the outer angle of the eye. This was caused by an injury received by a chip suddenly forced through the outer angle of the eye through the edge of the cornea. This destroyed a portion from the ligament from the margin.

NOW

This is our patient who had previously undergone cataract surgery (Patient #69) with an ancient method widely used called "couching." With this method, the cataract surgeon uses a long knife or needle to break the opacified lens into smaller pieces, thus facilitating its absorption and pushing the resulting pieces out of the visual axis into the vitreous cavity. In this case the patient has retained lens material which has not completely reabsorbed. These retained lens fragments may sometimes remain in the anterior chamber angle or in the posterior chamber for some time. Patients with retained lens fragments present with varying degrees of inflammation depending on the amount of lens material and the time elapsed since surgery. The clinical signs of retained lens material may include uveitis, elevated intraocular pressure, corneal edema, and vitreous opacities causing visual loss. Retained cortical lens material does not necessarily require surgical intervention since it is less inflammatory and more likely to reabsorb over time than nuclear material.

Observation is warranted for patients with small amounts of retained lens material in the hope that it will be reabsorbed. Inflammation should be controlled with corticosteroid and nonsteroidal anti-inflammatory drops as well as cycloplegics. Intraoccular pressure (IOP) can be controlled with topical agents or with carbonic anhydrase inhibitors given systemically. Surgical intervention may be necessary to remove residual lens material in the following situations:

- Presence of a large or visually significant amount of lens material
- Increased inflammation not readily controlled with topical medications
- Medically unresponsive elevated IOP resulting from the inflammation
- Associated retinal detachment or retinal tears
- Associated endophthalmitis

If the posterior capsule is intact, simple aspiration of the residual cortex through an anterior incision may be carried out with an irrigation/aspiration system. If there is a defect in the posterior capsule, pars plana vitrectomy and removal of lens material with the adjacent vitreous may be indicated. Chronic glaucoma and cystoid macular edema are more likely when intervention is delayed more than three weeks after the cataract surgery.

In this case, the patient has some degree of inflammation in the anterior chamber (synechiae between lens fragment and iris), damage to the angle, and cloudy vision. The cloudy vision could be secondary to inflammation, cystoid macular edema, post-operative astigmatism, corneal edema, or elevated intraocular pressure. This patient would most commonly be taken to the operating room for lens material removal and exploration as well as extensive topical medical therapy. Please also see discussion of Patient #76.

Dr. James Chodosh
Dr. Matthew F. Gardiner
Dr. Joan W. Miller
Dr. Sotiria Palioura
Dr. Athanasios Papakostas

Patient #84: Hand trauma

THEN

A 56-year-old male was admitted on May 26, 1823, two months after he fell on his left palm with all his weight. The pain started later in the day in his little finger and went up his arm to his elbow. It began to swell and had inflammation. On May 1 he had seen a doctor who relieved some of the pain. Poultices were used and swelling broke near the wrist with minimal discharge. It appeared the bones in the wrist were rubbing against each other. Blistered surface had a copious discharge. On June 19 the hand swelling had diminished.

NOW

Minor injury

Patient #85: Breast mass

THEN

A 32-year-old female was admitted on May 27, 1823, complaining of loss of appetite and headache. She had been sick for nine months with fever and cough. Her bowels were loose and inflamed and bled frequently. She had dropsy and she had been tapped times two, the last time about two months earlier. Now she complained of dizziness and weakness. She had stomach pain radiating to her shoulder. There was a tumor above the head of the sternum; there were hard swellings in the right breast, removed by surgery. She had a number of days of nausea and violent lower abdominal pain. She was transferred to Dr. Jackson.

NOW

A 32-year-old female patient with no past medical history was admitted to the surgical service with a several-month history of anorexia, headaches, fevers, and cough. The patient also notes that she has had diarrhea with blood mixed in the stool, and over the past few weeks has felt quite weak and dizzy. She also reports epigastric abdominal pain radiating to her left shoulder. She takes no regular medications and does not smoke tobacco nor consume alcohol. On examination, her temperature is 99.6 degrees, blood pressure 90/50, pulse 115 (sinus tachycardia), respiratory rate 22, and oxygen saturation 95% on 2 liters nasal cannula oxygen. She appears pale and ill, lying still on the stretcher. Sclerae are anicteric. In the right supraclavicular fossa near the head of the sternum is a 2 cm, firm, nontender lymph node. Chest is clear to auscultation bilaterally with decreased breath sounds at the bases. Abdomen is distended with a fluid wave, consistent with ascites. Breast exam shows a 4 cm, firm, mobile mass in the upper, outer quadrant of the right breast with retraction of the overlying skin. There are several firm, enlarged lymph nodes in the right axilla. Laboratory studies are notable for a hemoglobin of 7.5 gm/dL and a WBC 15, 000. Chest x-ray shows bilateral pleural effusions and no parenchymal infiltrates or nodules. CT scans of the chest, abdomen, and pelvis show bulky adenopathy in the right axilla and right supraclavicular fossa, bilateral pleural effusions, a 4 cm spiculated right breast mass, intra-abdominal ascites, innumerable bilobar liver metastases without biliary ductal dilation, and omental caking.

The patient was admitted with a diagnosis of presumed metastatic breast carcinoma. She was transfused two units of packed red blood cells, and follow-up hemoglobin was 9.5 gm/dL. An ultrasound-guided paracentesis by interventional radiology yielded three liters of yellow, clear ascitic fluid which was sent for cytology, and this returned as positive for malignant cells (adenocarcinoma), most consistent with a breast primary based on positive immunostaining for estrogen and progesterone receptors. A medical oncology consultation was obtained, and the patient was transferred to their service for consideration of palliative chemotherapy.

Dr. Barbara Lynn Smith

Patient #86: Pelvic mass

THEN

A 44-year-old married female was admitted May 6, 1823. She had some infection of the lungs for which she had been treated. Four years earlier she had had sores over the inside of the labia accompanied by discharge, fevers, and dysuria which continued for a long time. She had pain the back and a hard time in an erect position. The vaginal discharge had been present for some time and was often purulent. Sometimes there is fecal matter. The edge of the anus is very much inflamed and swollen but flabby above. Discharge of pus is not usual except on going to stool. The rectum appeared to be much diseased by cancer beyond the verge and she often has pain in the loins and lower part of the sacrum.

Genitals and uterus are perfectly healthy. The uterus was lower than natural. The mouth of the womb was not at all tender. It was smooth and moist and not swollen. Pressure on the rectum from the vagina produces severe pain. She complains of weakness, stomach pain, and loss of appetite. She was given some medication and six leeches every day. She complained of pain all over. On June 13, she had less pain around the pelvis and less difficulty in passing feces and appeared to be getting better. She complains of pain in the hips. She has not had leeches. But by June 23, she had a great deal of pain in the back, hips, vagina, and over the region of the sacrum. She had pain also in the bowels. She had more discharge. First, leaves of hemlock were applied. On July 1 she said she was somewhat improving and wished to be discharged. She was discharged on July 1, 1823.

NOW

A 44-year-old Ashkenazi Jewish female was seen December 17, 2012, in the Mass General Department of Obstetrics and Gynecology for non-specific pelvic symptoms.

Her medical history was notable for a severe primary herpes simplex infection four years earlier that led to a brief hospitalization. The event caused her to separate from her husband for six months, but they have since reconciled. She had no children, was on no medications, and had never had surgery. Her family history is significant for her mother who died from breast cancer at age 32, and a maternal aunt diagnosed in her mid-20s.

She described intermittent purulent vaginal discharge with particulate matter suspicious for stool and worsening rectal pressure. She noted progressive weakness, stomach pain, and loss of appetite. Curiously, she had been seen by three medical professionals over the past several months with stool softeners and other medications prescribed, but no one had performed a pelvic examination. Her abdomen had an obvious fluid wave and epigastric firmness. She was found to have a normal appearing cervix, but a fixed pelvic mass on bimanual examination that extended deeply into the rectovaginal septum and was eroding the posterior vaginal mucosa. A small fistulous connection was seen. Rectal examination revealed significant extrinsic tumor compression. She had 1+ bilateral pitting edema in her lower extremities.

She was sent for blood tests, imaging, and a gynecologic oncology consultation with a presumed diagnosis of advanced ovarian cancer. Her CA125 returned at 2550 (normal range, <30). The chest CT was normal, but abdominal-pelvic scan showed large volume ascites, omental caking, and a 15 cm pelvic mass. She was seen urgently on December 20 to be counseled about primary debulking. A research assistant also consented her for tumor, blood, and ascetic fluid collection at the time of surgery in order to facilitate preclinical drug testing of ovarian cancer. She was taken to the operating room on December 22 by faculty from gynecologic oncology. At the time of surgery, 6 liters of ascites were drained, the infracolic omentum is detached and sent for frozen section. The diagnosis of serous epithelial ovarian cancer was confirmed. An en bloc pelvic resection was performed to remove both ovaries, uterus, cervix, rectosigmoid and all encircling pelvic peritoneum with confluent tumor. The descending colon is mobilized and a low rectal anastomosis was performed. The operation resulted in no gross residual disease. An intraperitoneal (IP) catheter was placed.

Post-operatively, she had an uneventful recovery and was discharged home on December 29. On January 2, 2013, the patient's case was presented at the Multidisciplinary Gynecologic Oncology Tumor Board. Her pathology was reviewed before an assembled 30-person team of gynecologic oncologists, medical oncologists, radiologists, case management, radiation oncologists, nursing, Harvard Medical Students, Mass General OB/GYN residents and other trainees. Her care was discussed extensively and the known 16-month survival benefit of IP therapy versus standard IV paclitaxel and carboplatin was reviewed for this patient with optimally debulked stage IIIC ovarian cancer. An evidence-based consensus was achieved: participation in a collaborative group clinical trial incorporating IP chemotherapy. A phone call was made on the same day to review the results and plan with the patient.

She was seen on January 3 for surgical staple removal and discussion. She elected to enroll in the clinical trial after considering the alternatives and began treatment on January 10. Her CA125 declined precipitously and entered the normal range by February. During the course of her treatment, she was seen by a genetic counselor, opted for testing, and was confirmed to have an inherited germline BRCA1 mutation. She completed her sixth and final course of chemotherapy by the end of April and was declared to be in clinical remission. Due to the 80 to 90 percent risk of relapse, she would continue on the experimental study drug every three weeks indefinitely or until disease progression, as part of the trial protocol.

Gynecology Service

Patient #87: Leg ulcer/foot pain

THEN

A 17-year-old male was admitted on June 3, 1823. He had had a cold in October, 1821 and was confined to the house for a few days with fever. Immediately after this he experienced pain in his hips, which he called rheumatic pain. This kept him in his room for ten days longer and left his hip stiff and knee weak. In October, 1822, he had taken a voyage to North Carolina and caught another cold, which caused a recurrence of pain. He returned in the latter part of December, 1822. The pain was so severe it would not stop for five or six nights. He occasionally had chills and sweated profusely during the night. He slept in very few clothes. Any clothing made him sweat. He had pain in the knee. This pain extended from the groin to the greater trochanter and was usually dull. He had pain in the hip, particularly at night, and he could not lie on it. When he was perfectly still, he did not feel the pain. He was able to raise his foot and leg anteriorly without assistance. He had regular strength in the leg. He could not bend forward, his joint being so stiff.

On examination, the right leg was about a half inch shorter than the left, more flabby, and thinner, and the ankle about a quarter inch smaller. Also, the calf was 1 1/4 inches smaller, above the knee 2 1/2 inches smaller, and near the groin, one inch smaller. While he was in North Carolina, he was hurt on his shin. The skin was scraped off, which he did not notice until after he had returned. The wound was about the size of a 25-cent piece.

The hip and knee were bathed in spirits and the wound was dressed with multiple ointments. On June 10, after multiple days of treatment, the wound and skin sealed up but continued to look unhealthy. A poultice was applied. Blister was put on the hip which was very sore. The bowels were open. He said he felt better. A sprinkle of burnt alum was placed upon the wound and injured skin and dressed with a simple ointment. This was done three times per day. On June 26 he said the strength in his limb was improving. He was able to put his foot down and stand on it.

On July 2, it appeared from examination that he had contracted syphilis at some time. He had never been under a regular course of treatment and had many marks of it. Probably inflammation of the leg was due to it. On July 7 he was improving. By July 13, the ulcer on his leg was much worse, increasing in size. He kept getting plasters and poultices. Despite all these treatments he was no better. The ointment was omitted. By July 28 his ulcer had started to look well again and a caustic was applied. He was continued on the poultices and medication. By patient's request, he was discharged on October 1, "relieved."

NOW (See Patient #58)

Patient #88: Eye infection

THEN

A 26-year-old male was admitted on June 5, 1823 with soreness and increased discharge of mucus from his eye. Suppuration and inflammation continued to increase. Pressing on the anitcanthus of the eye, matter oozed out. At the time he said he had much pain and considerable swelling. The stricture of the duct appeared nearly, if not entirely, closed. The eye was irritated. He had gotten sawdust in it while at work. Zinc sulfate and aqua was inserted to wash the eye. Magnesium sulfate was also used. He made use of the wash and ointment and said his eye felt better. On June 8, his eye was in a good state for the operation, which was planned for the next day. The operation was begun with an incision just below the internal canthus and extended about four lines parallel with the same edge. The knife was then inserted into the sac and considerable pus followed. The probe was placed just about one inch into the duct, where it met with obstruction. The probe was withdrawn. The probe was again introduced and the obstruction was overcome and the probe was seen to pass into the nose. After a few minutes, the probe was withdrawn. A silver pin with a flat end was inserted into the wound, covered with a small piece of plaster. Post-operatively, at night, he was comfortable in every respect and wished to be discharged. The next day he had very little pain or soreness in the eye and was discharged with directions to change the pin every two or three weeks and to keep it in for about a year. During the first two weeks the head of the pin must be covered with sticking plaster to keep it in.

Note: There was some blockage between his nose and his eye which formed an abscess. It was approached via the duct of his eye into the nasopharynx or nose and then the pin was left there to let the fistula tract form better.

NOW

This 26-year-old patient, presenting with pain, soreness, swelling over the lacrimal sac in the lower lid medial canthal area, and discharge upon massaging the lacrimal sac, most likely suffered from dacryocystitis (i.e., inflammation of the lacrimal sac). The presence of obstruction in the lacrimal drainage system upon irrigation confirmed the diagnosis. Dacryocystitis can progress to a lacrimal sac abscess and to orbital or preseptal cellulitis. The patient's symptoms are classic for dacryocystitis and the initial work-up today would consist of a complete ocular examination and a CT scan of the orbit and paranasal sinuses to assess for preseptal or orbital cellulitis, especially if proptosis, restriction, and/or pain with eye movements were present. The discharge expressed from the punctum would also be cultured.

Today, probing and irrigation of the lacrimal drainage system are not indicated during an acute infection. A systemic antibiotic regimen that covers streptococcal and staphylococcal organisms is started first and is adjusted accordingly based on the culture results and the clinical response of the patient. Topical antibiotic drops can be used as an adjunct to the systemic antibiotics. If the condition is chronic and an abscess has developed, incision and drainage should be performed.

In the presence of an acute infection there is always a risk of progression to preseptal or orbital cellulitis and thus the patient is followed daily until improvement is documented. Once the acute episode has resolved and inflammation has subsided, surgical correction to reestablish patency of the lacrimal system is usually performed.

It is interesting to note that this patient's treatment involved the purposeful creation of fistula tract in order to facilitate drainage. This today is considered a complication of surgical therapy for such cases and is carefully avoided. Since antibiotics were not available, chronic drainage of the abscess and granulation by secondary intention was the only option.

Dr. James Chodosh
Dr. Matthew F. Gardiner
Dr. Joan W. Miller
Dr. Sotiria Palioura
Dr. Athanasios Papakostas

Patient #89: Syphilis

THEN

The patient is a 24-year-old married man admitted June 6, 1823. Before his marriage, he had had a venereal disease several times. He was treated about four years earlier and said that it appeared to entirely remove the complaint. He thought he was free of disease until a year ago when swelling in the ankle and testicles occurred. Sometimes the swelling in his testicles was very great, and other times the swelling subsided. The swelling mostly recurs when he has a cold and he would readily apply some mercurial ointment. He also had two ulcers, one inside the thigh and the other on the leg. These were now well. His ankle was now swollen. Motion in his joint was tolerable. He could use his limb with a crutch. Testicles were considerably enlarged, particularly the right which appeared to be about the size of a goose egg. He thought that this was the largest it had been. The soreness is very great. The testicles were rubbed with strong mercurial ointment twice a day. The testicles and pain in his ankle were less sore and inflamed. Blister around the ankle was discharging. Swelling in the testicle was diminishing in size. He had no bowel disturbance. His mouth was a little sore, bowels open. The swelling lessened. He received the pill at night. On June 23, he felt that his foot was more swollen but there was no pain except when he moved it. He was feeling tolerably well. Six leeches were ordered. He said the leeches did him good. This was done on July 2. Again, on July 7, ointment was omitted and leeches were applied. The patient wished to be discharged and was discharged on July 14, "much relieved."

NOW (See Patient #15)

Patient #90: Eyelid scar

THEN

A 26-year-old male from Hampton Falls, N.H. came into the hospital June 6, 1823 with a cancerous affliction at the root of the nose. In 1821, he had had a procedure causing enlargement, mostly confined to the lower eyelid below the internal canthus of the right eye. He had been to many physicians and had had various applications. For the last few months he had darting pains. Of late, however, he had been free of pain. He thought the general appearance had improved. On the diseased part, he discharged a thick matter about every week, formerly it broke more often. Each time it discharged, it amounted to only a few drops. Now the lower eyelid is drawn and is contracted with scars–cicatrization. On June 10, the scab fell off and the tumor looked better. The indurated part looked improved in appearance and the patient looked less cancerous. He was advised not to have an operation but he insisted on it. On June 12, the affliction on the eyelids exhibited no marks of a cancerous tendency. He experienced his lower eyelid being contracted and turned out towards the cheek. He was prepared for an operation to remedy the deformity of the eyelid.

For the operation, the patient was seated in a firm position with his head inclined and held against the back of the chair. The operator went under the eyelid from above downward and brought it directly out and forward making a division of the indurated portion. After which he carried the incision further and deeper with his scalpel and removed a triangular piece, cleaned the wound, and brought the edges together. He inserted two stitches, these being tied. Lint compresses were applied. On June 13, it was noted that since the operation the patient had felt very little pain. He says he feels well, bowels moving, appetite good, and deformity was less. He was discharged on June 16.

NOW

Though the case is labeled "eyelid scar," the account opens with a description of a "cancerous affliction at the root of the nose." The first paragraph tells the story of a lesion alternately draining and healing over the course of two years. Though basal cell carcinomas can linger and grow slowly over long periods of time, they should not completely regress and leave a scar as is described in this case. Therefore, we can assume that the "tumor" was a mass, though probably not a malignancy. Likely possibilities include a persistent chalazion of the left lower lid near the medial canthus or chronic dacryocystitis. Styes or chalazia rarely persist beyond several months and should not drain over a period of years. Dacryocystitis, which can occur in exactly the described location (medial canthus near the bridge of the nose) is an infection of the lacrimal sac and can be acute or chronic. The disease process is thought to begin with blockage of the normal tear drainage pathway, either from dacryoliths (acquired stones), scarring or dysfunctional valves in the sac and ducts. The ensuing stasis leads to colonization and infection with organisms such as staphylococcus or streptococcus. Patients can present acutely with fever, swelling, pain, and redness near the inner corner of the eye adjacent to the bridge of the nose. Cases can be chronic from intermittent obstruction and subsequent incompletely cleared infections which sometimes cause fistulous tracts to form to the facial skin (as appears to have been the case with this patient). Now, patients are treated with antibiotics (often intravenous) and/or surgical drainage with good success. For Patient #90, although it eventually cleared, he was left with "cicatrization" – scarring which led to his eyelid being chronically turned outward, otherwise known as a cicatricial ectropion.

The patient underwent excision of the scar and reconstruction of the defect by reapproximating the skin with sutures. Ectropion comes in several varieties, but some degree of lid laxity is involved in most. Lateral lid tightening procedures are commonly employed in surgical repairs, but cicatricial types have addition concerns in that anterior lamellar shortening must be addressed as well. This is usually done with skin grafting in order to relieve traction on the lower lid from scar tissue. Such procedures were not available to surgeons of the past and simple scar excision in our patient's case might only lead to development of further ectropion in the future.

Today, surgical procedures are done routinely for patients such as these with excellent success rate, thanks to the development of microsurgical techniques and grafting. Now, a patient such as this one would be treated as an outpatient under local anesthesia and hence would not require the ten-day in-house stay so described.

Dr. Joan W. Miller

Patient #91: Venereal disease

THEN

This was a 21-year-old male who contracted gonorrhea about three weeks prior to admission. This continued for about four days immediately after a small chancre came out on the glans penis. This was followed by two others over the skin of the penis, which were still smaller and more superficial. Chancres discharged very little and itched greatly. The penis was a little swollen. Plumbi compress, after the medication, was placed and the itching on the penis was gone. Three times a day, mercurial ointment was rubbed into the thighs, and the shankers were well dressed with simple ointment. By June 16, the sores were healing well. There was no burning with urination. His mouth was affected by the mercury and the sores readily seen. He was discharged on June 20, 1823.

NOW (See Patient #15)

Patient #92: Cataracts

THEN

The patient was a 61-year-old widow who came to the hospital June 10, 1823 with a cataract in each eye. The cataract in the left eye formed seven years earlier and in the right about eighteen months earlier. No peculiar symptoms were described at the commencement of their formation. Because of the cataracts, she had had slight injuries. The cataract in the left eye formed very rapidly and had recently altered very much in appearance. She said she could barely distinguish light from darkness. She was in better position than some others. The cataracts looked white.

Belladonna was given to both eyes the morning of the operation, June 12. Cataract knife and assorted instruments were assembled for the operation. The patient was fixed, her head held inclined against the chair. The left eye was done first. This was done by introducing the spear-pointed needle into the cornea and making a line from the circumference, and carried forward, punctured, incised the capsule of the lens. After the capsule was cut in several directions, the needle was withdrawn. The next part of the operation was the depression of the right eye and was done with the left hand of the operator. The same needle was used and introduced about two lines' breath from the circumference of cornea. The lens was detached, turned downwards, and depressed below the iris while the lower part remained attached. After the operation, dressings were applied. The patient appeared to suffer, but not very much. Soft compresses on the eyes with a tincture of opium were given. She slept part of the night. She was given laudanum. Pain continued. She had nausea and vomiting. Venesection was done with some lessening of pain. Medication was hydrangea and pulverized opium. Blisters were applied behind each ear, along with leeches. The leeches did her some good. She had taken mercury pills at night and in the morning.

Venesection was done from a large orifice but she soon felt faint. Her eyes were inflamed with the right eye being worse. Leeches to the right temple was the treatment used to get the swelling down around the eyes. Ointments, compresses, blisters, and leeches were used.

On July 2, she was still getting leeches and blisters. The eyes were less inflamed. Her mouth was sore. On September 2, she was seeing better. Inflammation continued. She was sometimes able to see light. On September 18 she underwent another operation. The cornea was transparent, the lens was opaque, and the pupils considerably enlarged. A cornea knife was used and an incision was made in the margin of the cornea. Incision pushed the lens with considerable force from the eye. On examination of the lens, it was found to be softer than in the former operation. It was too hard for the needle to penetrate and what was remarkable was that the capsule was diseased and nearly destroyed. Only a few shreds remained. This operation was very successful and gave more satisfaction because belladonna was applied. On October 1, a little inflammation was present in the eye. She could distinguish any person in the room. She got Lignum and plumbi acetate to the eye. Magnesium sulfate was given. Blisters and leeches were reapplied. She was discharged, by request, on November 17, "cured."

NOW

This is a case of a white, mature cataract in the left eye and possibly a hypermature cataract in the right eye with liquefied cortical material. The case report describes the right lens as softer; most probably secondary to the liquefied cortex but that the central core of the cataract was harder to penetrate. The capsule was wrinkled and shrunken because of liquefaction of the cortical material which sometimes leaks out of the capsular bag into the interior of the eye.

Pre-operatively, it is prudent to inquire about a history of trauma or previous ocular surgery that may have compromised the integrity of the lens capsule or zonules. It is important to note the presence or absence of a relative afferent pupillary defect, phacodonesis (lens instability), phacomorphic angle narrowing, or an inflammatory response to the hypermature lens. In virtually all such cases, the cataract will preclude the surgeon's view of the fundus. A B-scan ultrasound is warranted to rule out retinal detachment or other posterior segment pathology.

Modern cataract surgery is performed as an outpatient procedure with phacoemulsification techniques. Pre-operatively, the eye is anesthetized with either topical tetracaine or a retobulbar block containing lidocaine. The eye is dilated with phenylephrine and tropicamide and prepped with topical povidone iodine. A 15° sharp blade is used to create a small paracentesis, placed approximately two or three clock hours away from the site where an incision will be made for the phacoemlsification handpiece. A viscoelastic gel is then instilled to protect intraocular structures and allow more control during creation of the surgical incision. A clear cornea incision 2.6 mm wide is made at twelve o'clock, and then a continuous curvilinear capsulorrhexis is made often with Trypan blue stain (given the white nature of the cataract).

Subsequently, hydrodissection is performed to separate the peripheral cortex and lens from the capsule. Following hydrodissection, the surgeon emulsifies the lens by means of ultrasonic energy through the phacoemulsification handpiece. A second fine instrument (sometimes called a "cracker" or "chopper") is also used through the second incision to facilitate breaking the nucleus into smaller pieces for aspiration out of the eye. After phacoemulsification of the lens nucleus and cortex is complete, an artificial lens made of acrylic or silicone is implanted in the eye to maximize the vision. Post-operatively, the patient is given topical antibiotic and steroid drops and is seen again one day, one week, and one month after surgery. Please also see discussion of Patient #51.

Dr. James Chodosh
Dr. Matthew F. Gardiner
Dr. Joan W. Miller

Dr. Sotiria Palioura
Dr. Athanasios Papakostas

Patient #93: Cataracts

THEN

The patient was a 32-year-old physician from Maine. He was admitted on June 11, 1823. He had had a decrease in vision in his right eye for nearly five years, and the vision had been decreasing gradually. One year earlier, the left eye also became affected in the same manner but had grown worse more rapidly than the right. Objects could not be distinctly observed. On attempting to read letters, they appear blended together, often receding and changing place. Small objects could not be perceived and large ones were ill-defined. Vision in both eyes was about the same; capacity was not particularly remarkable but greater in the right than the left. Opacity of the lens and capsule was perfectly clear and susceptible to light. Pupils easily dilated. Application of belladonna to each eye pre-operatively.

The pre-operative preparation was the same. Pupils were thoroughly dilated by belladonna, and soft bandages, compresses, needles, and knives were all assembled. Operation was first on the left eye. A needle was entered aligned behind the cornea. An assistant raised the upper lid with his finger and the operator, with the forefinger of his left hand drew the lid, then the needle was passed directly into the globe of the eye. The needle was turned near the internal margin of the iris. Using the edge of the needle backwards, moved it in different directions until the lens was well divided and firmly brought forward into the anterior chamber. As fragments were created, the needle was withdrawn and soft pressure was applied. Subsequently, the operation was performed on the right eye. It was operated on in the same way. The eyes were kept wet with compresses. He did not have much pain. It was not until June 18 that his cataracts were quickly dissolving. On July 6, it was noted that his cataracts were quickly dissolving and a good part of the pupils appeared transparent. He was discharged on July 7.

NOW

This was a 32-year-old patient who presented in 1823 with a gradual decrease in vision in both eyes. He was unable to distinguish objects and reported that "on attempting to read letters, they appear blended together, often receding and changing place." The patient underwent cataract surgery in both eyes and was discharged home shortly thereafter. Unfortunately, we do not have information on his vision after the surgery.

Although cataracts can cause blurry and hazy vision, given his young age and symptoms of metamorphopsia (i.e., distortion of vision) in both eyes, today's work-up would include a careful retinal examination. Macular disease such as central serous chorioretinopathy, macular edema, choroidal neovascular membranes (sometimes due to ocular infections from histoplasmosis), and bilateral serous retinal detachments associated with optic disc pits can cause symptoms of vision distortion. Since central serous chorioretinopathy is usually seen in men aged 25 to 50 years and is associated with a type A personality (patient is a physician), it is a possible diagnosis in this patient. Of course, the most important part of the ophthalmic examination that would lead to such a diagnosis, that is, a dilated fundoscopic examination of the retina, was not feasible in 1823. The direct ophthalmoscope became available in the 1850s and the use of condensing lenses for indirect ophthalmoscopy emerged more than a century after that.

Today, imaging techniques, such as fluorescein angiography of the retina and optical coherence tomography of the macula, aid in confirming the diagnosis. Currently, there is no definitive treatment for central serous chorioretinopathy and visual prognosis is worse for the small subset of patients who develop recurrent disease or those who have multiple areas of detachment. However, fast, spontaneous resolution of the serous detachment may occur with some patients resulting in visual recovery of at least 20/30 and the majority of patients generally regain reasonable levels of acuity without treatment. Follow-up is recommended every three months until resolution of the serous detachment is complete. Photodynamic therapy, a drug/laser combination developed for age-related macular degeneration, is often used for chronic cases with some success. However, this is one disease where we have advanced little beyond our colleagues of two centuries ago. Perhaps our descendants in 200 years will look back upon us with the same quizzical amazement that we have such limited understanding of so simple a problem.

Dr. James Chodosh
Dr. Matthew F. Gardiner
Dr. Joan W. Miller

Dr. Sotiria Palioura
Dr. Athanasios Papakostas

Patient #94: Carcinoma of the rectum

THEN

A male patient from Salem entered June 13, 1823, with a fungal tumor of the rectum and was advised to return the same day, the disease being incurable. He was discharged, "incurable."

NOW

This poor fellow probably had an adenocarcinoma of the rectum, or perhaps a squamous cell carcinoma of the anus. Far from being incurable today, rectal cancer is often successfully treated by radical resection (frequently sphincter saving), sometimes with combined pre-operative radiation and chemotherapy, as well as post-operative chemotherapy. Anal carcinoma his frequently cured by chemotherapy and radiation alone.

Dr. Paul C. Shellito

Patient #95: Soft tissue injury

THEN

A 7-year-old girl came into the hospital on June 19, 1823. About three weeks earlier she received an injury to the right knee by putting the knee between a tree and a large stone, which lacerated the skin extensively and denoted the ligaments and tendonous substance. The injury was over and above the external condoyle. The injury was about the size of a hand; the internal condoyle was much smaller. There was no dislocation of the bone. The joint was movable. The knee was considerably swollen. The skin was drawn together and secured by an interrupted suture and then started to separate and the skin started to slough. The wound on both sides was dressed with small pieces of lint covered with ointment. On July 1 the wound was improving. On July 7, scarring was noted to be going on. Patient started to suffer on July 30 with unexplained severe abdominal pains, indicating a great deal of suffering. Opium was given. The prescription was repeated. The pain at night was relieved with one dose of opium and remained unexplained. The wound to the knee was healing well and she was discharged on September 12.

NOW (See Patient #58)

Patient #96: Soft tissue inflammation

THEN

A 25-year-old male from Ireland came into the hospital on June 21, 1823 having had a year of pain and swelling in the knee without any known cause. This continued for a time and finally abated so that it was tolerably strong. However, upon exercising the limb thoroughly a month after the inflammation, the pain again occurred and continued to the time of admission. Doctors did not know what was wrong with his knee. They put poultices on it. They created blisters around it. He complained of dysuria and terminal hematuria. On July 10, the knee continued to discharge copiously. He had a stoppage of water and no history of strictures. He was discharged on September 15, on poultices, much relieved.

NOW

Minor orthopedic condition

Patient #97: Varicocele

THEN

A 28-year-old male was admitted to the hospital on July 2, 1823. He said that for 15 years he had been troubled with varicocele and pain in the scrotum, particularly on the left side. The scrotum was much enlarged and hung down. He was flaccid and wasted. He had considerable pain, occasionally through his limbs and body, which was aggravated by exercise, more severe on the left. They applied leeches to his scrotum. The swelling on the scrotum initially was diminished by the leeches. On July 7, he was much the same. He was told he could be discharged if he wished and to pursue the course recommended. He was given some pills to take at night and was discharged "relieved."

NOW

In current times, this patient would have more likely been identified earlier by his pediatrician or primary care physician. In addition, the patient would be evaluated as an outpatient, and not admitted to the hospital as he was in 1823. The patient would be examined to determine testicular size and the grade of the varicocele, but also to assess if it decompresses when lying down. If it does not, imaging to assess for a retroperitoneal or abdominal mass should be performed. Cross-sectional imaging of the retroperitoneum is indicated if a varicocele is isolated to the right side, if the right side is larger than the left in cases of bilateral varicoceles, or in circumstances of varicoceles that do not decompress when supine.

In this particular case, given that this patient was reported as being flaccid and wasted, imaging with an abdominal and pelvic CT scan would be prudent even if the varicocele does decompress. In many cases today, the patient would be evaluated with a color Doppler scrotal ultrasound. This would assess the degree of varicocele and would assess for size and volume of the testes. However, if a varicocele is easily examined and graded on physical exam, an ultrasound may be skipped and testicular volume can be directly estimated with an orchidometer. Identification of testicular hypotrophy, or loss of normal volume, is important, as this finding is suggestive of spermatogenic dysfunction and may be associated with infertility, with the varicocele being a causative and potentially correctable factor. In addition, a semen analysis and often serum follicle-stimulating hormone (FSH) and testosterone are frequently obtained to help predict fertility potential and are helpful in decision-making regarding surgical treatment vs. observation of the varicocele. Although observation is the preferred option in the majority of adolescents with a varicocele, indications for surgery include significant testicular hypotrophy, associated pain, or abnormal semen analysis. Leech therapy, as was used in this patient in 1823, is not used in modern times. Surgical varicocele repair would be performed via a subinguinal or inguinal microsurgical approach with an operating microscope to aid in sparing lymphatics and the spermatic artery. A laparoscopic varicocelectomy would also be an option. Surgical repair would typically be performed as outpatient ambulatory surgery. One must be cautious in using pain as an indication for surgery, as surgery can fail to cure the symptoms of pain in up to 30 percent of cases. Fortunately, in a modern health care system, pain is reported as a symptom in less than ten percent of varicoceles.

Dr. Adam Feldman

Patient #98: Varicocele

THEN

A male patient who was a student of medicine, approximately aged 21, entered the hospital July 12, 1823 with a varicocele. For about a week he had been affected with pain about the hip and loins. He had a sensation of fatigue when standing upright and sitting; however, it was barely perceptible when lying down. Warm fermentations were placed on the scrotum. On July 13, the pain in the hip was much the same. He took medication to regulate his bowels and used blister to draw swelling, getting some relief from the pain. He had no inconvenience from the varicocele and was treated with ointment. He was able to walk about on July 15. He applied poultice and powder in the morning. On July 19 he was quite sore. He was discharged "cured."

NOW

A 21-year-old presenting in current times for evaluation of scrotal pain may be found to have a varicocele on physical examination. Varicoceles are present in approximately fifteen percent of the general male population but are more common in men with infertility, approximately one third of whom will have a varicocele detected. Varicocele-associated orchialgia is generally described as a dull, aching heaviness in the scrotum, worse with standing and relieved by lying down, as was the circumstance for this patient. The goal of varicocele repair is to optimize fertility potential, not specifically to accomplish pain resolution, as data regarding improvement or resolution of orchialgia in the setting of varicocele repair are limited.

In contrast to the eight-day hospital stay this patient experienced in 1823, the evaluation and treatment of varicocele would be performed on an outpatient basis in the modern era. Diagnosis is accomplished by a thorough physical examination in the office. In some circumstances a scrotal sonogram may be performed. If varicocele repair is indicated based on ipsilateral testis hypotrophy or abnormal semen parameters, this can be achieved via surgery or percutaneous embolization. Surgical varicocelectomy can be undertaken through a variety of approaches: retroperitoneal, laparoscopic, microsurgical subinguinal, or inguinal.

Optical magnification with operating loupes or a microscope allows the surgeon to identify all of the veins within the spermatic cord that need to be transected, while preserving the arterial supply and lymphatic return. No degree of fermentations, poultices, or powders will fix a varicocele! Each of these procedures to repair a varicocele takes no longer than one hour and is performed with the patient under general anesthesia or sedation. The patient is discharged with instructions to avoid strenuous activity for a short period post-operatively.

Dr. Cigdem Tanrikut

Patient #99: Syphilis

THEN

The patient was a 22-year-old laborer admitted July 19, 1823. It had been a month since he had contracted syphilis. He said he had never had the disease before. He had been under no medical treatment. He now has phimosis and copious drainage accompanied by dysuria. Sometimes he has been unable to pass water. The penis is very much inflamed and swollen. Plumbi lead ointment was applied to the penis by means of a white cloth. They put leeches on his penis to reduce the swelling, three times a day. The discharge from the penis started to decrease. On July 23, he had very little discharge, however he had discharge from his urethra by July 26. Mercurial ointment was used. They looked for mouth irritation and could not find it but by July 30 his mouth was sore. The penis was dressed with compresses and the chancre was much reduced and continued to improve. An operation was performed on August 13 for phimosis. Instruments were placed under the foreskin and initially slit up a line with an X to allow the skin over the glans on the right side to flap so that there was essentially a dorsal slit. Lint was applied. He was discharged "cured."

NOW (See Patient #15)

Patient #100: Hemorrhoids

THEN

The patient was an apothecary from Salem, age 27, who came into the hospital July 21, 1823 with a history of piles. Previous to the piles he was also very constipated and had gone ten days without a stool. He had been taking a lot of physics. The piles had bled very little for the last six months and he had a lot of pain in the stool. He was given laxatives. The hemorrhoids were not protruding. An operation was prepared. The instruments were armed and unarmed needles, scalpel, tenaculum, compresses, bandages, pins, sponges, and wine and water.

The patient was inclined over the bed and the parts well examined. The physician began the operation. They commenced by taking up the hemorrhoid position with the tenaculum and removing it with the scalpel. In this manner, he went after the verge of the anus, cutting away three different portions. Afterward, he cut away the projecting part from the center, which completed the operation. Bleeding was not very great. The patient got a tincture of opium, 30 drops. The concern was getting him to defecate so he was given laudanum with some relief. He had a constant desire to pass water without the ability to urinate. He was now in pain, which was being controlled with opiates. He still could not urinate. Fomentations were placed on the anus. He did not require medication. He had a bowel movement without laxative. Two pieces of hemorrhoid were left, which required the knife. Two projecting pieces of hemorrhoid were removed with the tenaculum and knife and the usual dressing was applied. They used oil to lubricate his bowels. The parts operated on looked well.

Gradually he became less swollen, his bowels were open. Stools were of a good size. On August 10, he was doing well. On August 15, the wound looks totally well. He was discharged on August 23 with instructions to wash parts with warm water twice a day. He was discharged "cured" on September 2, 1823.

NOW

The same comments apply for this case as for Patient #48, except that apparently this patient at least had his surgery done in the appropriate prone position. A second operation to remove residual hemorrhoid tissue was required 26 days later. That would be very unusual today, as would be his 43-day hospitalization.

Dr. Paul C. Shellito

Acknowledgments

I wish to credit Dr. Paul S. Russell and the History Committee of the Massachusetts General Hospital for creating the "Then and Now" concept. I would like to express my gratitude for Dr. Russell's support and encouragement. Jeff Mifflin, our archivist, deserves to be recognized and thanked for locating and carefully handling the archival material used for the "Then" sections of this manuscript. I also wish to thank the MGH's Chiefs of service and the Chief of Ophthalmology at the Massachusetts Eye and Ear Infirmary for encouraging their residents and fellows to create and prepare the high quality "Now" sections of this manuscript. Dr. Harry E. Rubash and Ms. Aimee Leyden of the Orthopedic service must be acknowledged for their enthusiastic support and organizational skills. Finally, many thanks to Arch MacInnes for his skillful design of the book cover, and to Marci Lindsay of Grey Area Editing for her meticulous attention to detail, which improved the quality of the manuscript immeasurably.